Website: Tglang.com

ISBN-9781481892193

Printed in the United States of America

The Pennsylvania State University

The Graduate School

Department of Aerospace Engineering

A GENERALIZED ENGINEERING DESIGN PROCEDURE

A thesis in

Aerospace Engineering

by

Thomas G. Lang

Submitted in partial fulfillment
of the requirements
for the degree of

Doctor of Philosophy

June 1968

Approved:

May 9, 1968

George F. Wislicenus.
Thesis Adviser and Head of the
Department of Aerospace Engineering

May 9, 1968

A.J. Sevik
Associate Professor of Aerospace
Engineering

ACKNOWLEDGEMENTS

The author wishes to especially thank Professor George F. Wislicenus for his general guidance, many important and basic suggestions, and careful review of the concepts generated during this study. Professor David P. Hoult is thanked for making certain key suggestions; Professor Maurice Sevik is thanked for his many suggestions and comments on fundamental engineering design concepts; and Professor Thomas C. Benton is thanked for his suggestions, careful review of the manuscript, and discussions relating to mathematical aspects. The writer also thanks Professor Barnes W. McCormick for his general comments and discussions on design theory and aerodynamics; Professor James W. Bond for discussions relating to the use of mathematical techniques, scaling, and possible techniques for extending the mathematical framework; and Professor J. William Holl for his review and discussions on cavitation.

The work of the author was supported by the U. S. Naval Ordnance Systems Command and the Naval Undersea Warfare Center as a part of the graduate study program in Background Research in the Field of Naval Hydrodynamics in the Department of Aerospace Engineering and the Ordnance Research Laboratory.

TABLE OF CONTENTS

Page

ACKNOWLEDGEMENTS.................................... ii

LIST OF TABLES vi

LIST OF FIGURES................................... vii

LIST OF SYMBOLS.................................... x

LIST OF DEFINITIONS...............................xix

Chapter

   I. INTRODUCTION.................................. 1

      Nature of the Study................................. 1
      Goal of Engineering Design.......................... 2
      Background.. 3
      Statement of the Problem............................ 7
      Outline of the Text................................. 8

  II. DEVELOPMENT OF THE DESIGN PROCEDURE.....................10

      Design Problem.....................................10
      Design Mission.....................................11
      Generalized Design Mission.........................15
      Mission Space......................................16
      Design Space.......................................17
      Mapping..18
      Regions of Mission Space...........................20
      Typical Design Forms...............................21
      Region Boundaries..................................22
      Mapping from Mission Space to Design Space..........26
      Dimensionalizing the Design........................28
      Classification.....................................30
      Mathematical Representation.........................30

 III. DESCRIPTION OF THE DESIGN PROCEDURE.....................39

      Outline of the Design Procedure....................39
      Advantages of the Design Procedure.................43
      The Design of Circular Tubes Subjected
           to External Pressure........................44
      The Design of Cylindrical Columns
           Loaded in Compression.......................60
      Economic Example...................................64

Chapter

IV.   DISCUSSION OF THE DESIGN PROCEDURE..................... 70

     Methods for Nondimensionalizing Variables........... 70
     Selection of Mission Parameters..................... 73
     Mission Parameter Ranges............................ 79
     The Number and Type of Design Form Solutions....... 80
     Regions, Boundaries, and Design Form Families...... 82
     Scaling... 83
     Optimized Scaling................................... 89
     Application of the Design Procedure to Research..... 91
     Future Engineering Design Theories.................. 98

V.   DESIGN OF SUBMERGED VEHICLES...........................102

     Generalized Mission.................................102
     Possible Design Forms..............................107
     Physical Relationships.............................107
     Selection of the Mission Parameters................115
     Status of the Problem..............................117
     Power- and Energy-Limited Vehicles.................118
     Power-, Energy-, and Density-Limited Vehicles......121
     Power-, Energy-, Density-, and Weight-
        or Volume-Limited Vehicles.....................124
     Classification of Submerged Vehicles...............125
     Numerical Examples.................................127
     Effect of Technological Improvements on the
        Performance and Form of Submerged Vehicles......130

VI.   SUMMARY AND CONCLUSIONS................................144

     Summary..144
     Conclusions..147

BIBLIOGRAPHY...149

APPENDICES...152

A.   DESIGN OF LOW-SPEED AIRPLANE WINGS AND
     NON-CAVITATING HYDROFOILS..........................152

     Specification of the Subdesign Problem..............152
     Generalized Design Mission of the Lifting
        Surface Problem................................154
     Possible Design Forms..............................155
     Physical Relationships.............................157
     Mission Parameters and Design Parameters...........165
     Design Equations...................................165
     Viscosity-Limited Optimized Lifting Surfaces.......166
     Viscosity- and Strength-Limited
        Lifting Surfaces...............................169

Appendices

Viscosity-, Strength-, and Cavitation-
Limited Hydrofoil Designs...........................179
Design of Airplane Wings and Hydrofoils
Which are Thickness Limited in Addition
to Being Viscosity and Strength Limited...........184
Design of Elasticity-, Cavitation-, Viscosity-,
and Strength-Limited Hydrofoils...................187

B. THE DESIGN OF HYDROFOIL CROSS SECTIONS...................199

General Characteristics of Hydrofoils.................199
Specification of the Generalized Design Mission.......201
Possible Design Forms.................................203
Physical Relationships................................203
Mission and Design Parameters.........................206
Selection of Subspaces of Mission Space
for Mapping.......................................207
Mapping from Subspace (a), ($C_L = 0$, $M' = 0$,
σ variable)..208
Mapping from Subspace (b), ($C_L = 0$, M' variable,
σ variable)..209
Mapping from Subspace (c), (C_L variable,
$M' = 0$, σ variable)................................223
Mapping from Subspace (d), (C_L variable,
M' variable, $\sigma = 0$)................................237
Means for Increasing the Leading Edge Strength
of Supercavitating Hydrofoils.....................245
Comparison of the Lift-to-Drag Ratios of
Supercavitating Hydrofoils Operating at $\sigma = 0$.....246
Mapping from Subspace (e), (C_L variable,
$M' = 0.0005$, σ variable).........................247
Mapping from Subspace (f), (C_L, M' and
σ are variable)....................................259
Transformation of the Three-Dimensional Subspace
of Mission Space into a One-Dimensional Subspace..262
General Comments on the Design of
Hydrofoil Cross Sections..........................266

C. GROUP THEORY AND DESIGN FORM TRANSFORMATIONS.............271

Transformation of Hydrofoil Cross Sections............271
General Design Form Transformations...................275

D. THE EFFECT OF SWEEPBACK ON THE INCIPIENT
CAVITATION NUMBER OF HYDROFOILS.......................278

E. CAVITY DRAG COEFFICIENTS FOR HYDROFOIL CROSS SECTIONS
WHICH CONSIST OF TRUNCATED ELLIPSES...................281

LIST OF TABLES

Table   Page

1. VALUES OF $y_o'(x')$ FOR THE NACA a = 1.0 (UNIFORM
        PRESSURE) MEANLINE AT C_L = 1.0.....................226

2. APPROXIMATE VALUES OF $y_1'(x')$ THROUGH $y_5'(x')$ FOR THE
        BASIC 2-TERM CAMBER, δ-THICKNESS, AND PARABOLIC
        THICKNESS DISTRIBUTIONS DESIGNED FOR σ = 0 AND
        INFINITE DEPTH................................232

3. L/D RATIOS OF VARIOUS SUPERCAVITATING HYDROFOIL
        FORMS AT σ = 0...............................246

4. REGION BOUNDARIES FOR SUBSPACE (f) AS
        A FUNCTION OF K..................................264

5. HYDROFOIL FORM CHARACTERISTICS CORRESPONDING
        TO SUBSPACE (f)..................................265

6. HYDROFOIL DRAG COEFFICIENTS FOR THE FORMS
        CORRESPONDING TO SUBSPACE (f)....................264

LIST OF FIGURES

Figure   Page

1. Schematic illustration of the mapping process........... 19

2. Illustration of a boundary in three-dimensional
   space... 25

3. Illustration of a two-dimensional mapping from a
   mission space consisting of two regions.............. 29

4. Design of circular tubes subjected to external
   pressure (p/f_p, p/E) space......................... 53

5. Design of circular tubes, $\left[p/f_p, \ 1.54 \ (p/E)^{1/3}\right]$ space..... 54

6. Design of circular tubes, p/f_p space.................. 56

7. Design of circular tubes, (p/f_p, f_p/E) space........... 57

8. Design of circular tubes, (p/f_p, f_p/E, e/R) space,
   $e/R = 0.005$... 59

9. Design of cylindrical columns loaded in compression..... 63

10. Generalized economic problem No. 1..................... 67

11. Generalized economic problem No. 2..................... 69

12. Lift coefficients and flow regions of a hydrofoil
    as a function of σ................................. 95

13. Operating range of a hydrofoil.......................... 97

14. Required hydrofoil operating range..................... 97

15. Possible forms of submerged vehicles................... 108

16. Vehicle limited by power or energy..................... 119

17. Vehicle limited by power and energy.................... 120

18. Vehicle limited by power, energy, and volume........... 122

19. Vehicle limited by power, energy, and density......... 123

Figure   Page

20. Improvement factor required for a speed increase
     of 25%..134

21. Effect of technological improvements on increasing
     the range..141

22. Possible planforms, thickness tapers, and cross-
     sectional shapes of lifting surfaces.....................156

23. Examples of lifting surface applications.................156

24. Viscosity- and strength-limited lifting surfaces........174

25. Optimum spanwise t/c variation...........................178

26. Viscosity-, strength-, and cavitation-limited
     fully-wetted hydrofoils..................................182

27. Aspect ratio and thickness-to-chord ratio of
     optimized fully-wetted hydrofoils........................183

28. Lift-to-drag ratio of optimized fully-wetted
     hydrofoils...185

29. Divergence limits of optimized fully-wetted
     hydrofoils...193

30. Flutter limits of optimized fully-wetted
     hydrofoils...198

31. Typical hydrofoil forms..................................204

32. Section modulus coefficient of truncated
     elliptical struts..219

33. Hydrofoil struts and drag coefficients mapped
     from Subspace (b)..222

34. Hydrofoil forms and drag coefficients mapped
     from Subspace (c)..236

35. Hydrofoil forms and drag coefficients mapped
     from Subspace (d)..244

36. Illustration of conditions imposed on the mapping
     from Subspace (e)..251

37. Hydrofoil forms and drag coefficients mapped from
     Subspace (e)...258

Figure   Page

38.   Boundaries of Regions I through IIe in
          three-dimensional space................................260

39.   Illustration of a three-dimensional mapping
          into hydrofoil forms..............................261

40.   The relationship of the different hydrofoil
          families corresponding to Subspace (f).............263

41.   One-dimensional representation of hydrofoil
          design form characteristics........................267

42.   Drag coefficients and physical properties of
          truncated ellipses................................284

LIST OF SYMBOLS

a	Speed of sound in a fluid (LT^{-1})
a_o	Value of C_{L_α} for a lifting surface which has an infinite aspect ratio
A	Cross-sectional area of a column (L^2); hydrofoil or airfoil planform area (L^2)
A_r	Aspect ratio = b/c
A_1	Tooling cost ($)
A_2	Cost of material per item ($)
A_3	Fabrication cost per item ($)
b	Hydrofoil span (L); characteristic thickness (L)
c	Chordlength of a hydrofoil (L)
c_k	Mission criteria, k = 1, 2, ..., r
C	Set of mission criteria
C_d	Hydrofoil or airfoil drag coefficient = $D/\frac{1}{2}\rho U^2 bc$; vehicle drag coefficient = $D/\frac{1}{2}\rho U^2 V^{2/3}$
C_{dc}	Cavity drag coefficient
C_{df}	Frictional drag coefficient of a hydrofoil
C_{di}	Induced drag coefficient of a lifting surface
C_{do}	Cavity drag coefficient at $\sigma = 0$
C_{dp}	Profile drag coefficient of a lifting surface
C_h	Ratio of the weight of a hydrofoil to the weight of an equivalent solid hydrofoil

c_i Correction factor for the induced drag

c_L Lift coefficient $= L/\frac{1}{2}\rho U^2 bc$

c_{L_α} $\partial C_L/\partial\alpha$

c_{Lo} Lift coefficient at $\sigma = 0$, $C_{Lo} = C_L - 2\sigma$

$(C_L)_{a = 1.0}$ Design lift coefficient of an NACA $a = 1.0$ meanline

c_M Hydrofoil moment coefficient = hydrodynamic moment about the midchord point $/\frac{1}{2}\rho AcU^2$

c_p Pressure coefficient = (static pressure at some point $-P)/\frac{1}{2}\rho U^2$

c_s Structural weight coefficient

c_t Torsional stiffness coefficient $\doteq 0.30$ for thin rectangular cross sections

c_v Volume of structural material in a lifting surface divided by bct

c_α Torsional mass moment of inertia coefficient $\doteq 2\pi$ for solid ellipse-like cross sections of hydrofoils

c_1 Section modulus coefficient $= 2I/t^3c$

c_2 Ratio of root thickness-to-chordlength to the mean thickness-to-chordlength

c_3 Ratio of root chordlength to the mean chordlength

c_4 Distance from the root of a lifting surface to the semispan center of pressure divided by b/2

c_5 Numerical constant appearing in Equations 174 and 175

d Maximum body diameter (L)

d_j Design form parameters, $j = 1, 2, \ldots, q$

D Hydrofoil drag (F); set of design form parameters; total vehicle drag (F)

D_a Drag of a vehicle minus D_ℓ (F)

D_c	Cavity drag of a hydrofoil (F)
D_ℓ	Drag of a lifting surface (F)
e	Elliptical out-of-roundness of a tube, measured as the maximum deflection from the desired circle (L); distance from the chordwise center of pressure of a lifting surface to the elastic axis (L)
E	Modulus of elasticity (FL^{-2})
E(k)	Complete elliptic integral of the second kind
f	Design bending stress including load factor and factor of safety (FL^{-2})
f_j	A function of a set of mission parameters
f_p	Proportional stress limit of structural material in compression (FL^{-2})
f_1, f_2, ...	Symbols used to denote functions
F	Froude number $= U/\sqrt{gc}$, $U/\sqrt{g\ell}$
g	Acceleration of gravity (LT^{-2})
g_q	A transformation from one design form to another within the same family
g_1, g_2, ...	Symbols used to denote functions
G	Modulus of rigidity (FL^{-2})
I	Area moment of inertia (L^4)
I_α	Mass moment of inertia of a hydrofoil about the elastic axis (FLT^2)
J	Torsional stiffness divided by $G = C_t ct^3$ (L^4)
k	Designates amount of camber of a 2-term hydrofoil camber line; a term used in the pi theorem; variable in an elliptic integral
K	Hydrofoil classification parameter $= (\sigma - C_L/2)/\sqrt{M'} = \sigma_o/\sqrt{M'}$

$K(k)$	Complete elliptic integral of the first kind
K_1	A nondimensional parameter defined by Equation 149
K_2	A nondimensional parameter defined by Equation 178
K_3	A nondimensional parameter defined by Equation 187
ℓ	Length of a column (L); wetted length of a ship hull (L); characteristic length (L); body length (L)
ℓ_c	Length of a cavity (L)
L	Hydrofoil lift (F)
m_i	Mission parameters $i = 1, 2, \ldots, p$
m'	Ratio of hydrofoil mass to the transverse added mass of the fluid
M	Applied bending moment about some cross section of a lifting surface (FL); set of mission parameters
M'	M/fc^3
n	Column support conditions at ends
N	Number of items sold
p	Pressure difference across a tube wall times the safety factor (FL^{-2}); nondimensional gross profit
P	Free-stream pressure (FL^{-2}); static pressure at the minimum operating depth of a submerged vehicle (FL^{-2})
P_ℓ	Pressure on the lower surface of a hydrofoil (FL^{-2})
P_u	Pressure on the upper surface of a hydrofoil (FL^{-2})
P_v	Vapor pressure of the fluid (FL^{-2})
P_1	Minimum pressure on a hydrofoil (FL^{-2})
q_d	Dynamic pressure at which divergence first occurs (FL^{-2})
q_f	Dynamic pressure at which flutter first occurs (FL^{-2})

q_ℓ	Optimization parameters, $\ell = 1, 2, ..., s$
Q	Optimization criterion
r	Radius of gyration of a column cross section (L), characteristic roughness height (L)
r_j	A function of mission parameters
r'	Nondimensional roughness height
R	Mean tube radius (L); vehicle range (L)
R_e	Reynolds number $= Uc/\nu$, $U\ell/\nu$
s_j	A function of mission parameters
t	Mean maximum thickness of a hydrofoil or lifting surface (L); tube wall thickness (L)
$t(x)$	Thickness distribution of a hydrofoil (L)
t'	Maximum hydrofoil thickness divided by $c = t/c$
$\bar{t}$	Local thickness of a hydrofoil divided by t
t_c	Maximum thickness of a cavity (L)
$t_o(x)$	Thickness distribution per unit maximum thickness
u	Circulation velocity around a hydrofoil (LT^{-1})
u_o	Velocity reduction on the lower surface of a supercavitating hydrofoil at $\sigma = 0$ due to lift (LT^{-1})
u_t	Velocity increase along a hydrofoil surface due to thickness (LT^{-1})
U	Free stream velocity (LT^{-1}); vehicle speed (LT^{-1})
U_f	Speed at which flutter first begins (LT^{-1})
U_ℓ	Average velocity along the lower surface of a hydrofoil (LT^{-1})
U_u	Average velocity along the upper surface of a hydrofoil (LT^{-1})
U_1	Velocity at the minimum pressure point on a hydrofoil (LT^{-1})

V	Volume of fluid displaced by a vehicle (L^3)
V_b	Volume of the buoyancy source (L^3)
V_d	Volume of drag reduction equipment (L^3)
V_e	Volume of energy-dependent vehicle components (L^3)
V_o	Volume of vehicle components which are independent of speed, range, or buoyancy requirements (L^3)
V_p	Volume of power-dependent vehicle components (L^3)
W	Load on a column (F); total vehicle weight (F)
W_a	Weight of a vehicle minus the structural weight of the lifting surface (F)
W_b	Weight of the buoyancy source (F)
W_e	Weight of the energy-dependent vehicle components (F)
W_o	Weight of the vehicle components which are independent of speed, range, and buoyancy requirements (F); weight of an airplane less structural weight (F)
W_p	Weight of power-dependent vehicle components (F)
W_s	Structural weight of an airplane wing (F)
W_{sa}	Structural weight of an airplane (F)
W_x	The weight of all components placed in or on the lifting surface (F)
x	Chordwise distance to a specific point on a hydrofoil from its leading edge (L)
x'	Ratio of x to c
y	Distance from the chordline to a specific point on a hydrofoil surface (L)
$y(x)$	Meanline distribution of a hydrofoil (L)
y'	Ratio of y to c
y'_e	Local semithickness of an ellipse divided by the length of the ellipse

y'_ℓ	Local nondimensional lower surface height above the chordline
y'_m	Local nondimensional meanline height above the chordline
$y_o(x)$	Meanline distribution for a unit lift coefficient (L)
y'_o	Nondimensional height of the NACA uniform pressure meanline for $C_L = 1.0$
y'_u	Local nondimensional upper surface height above the chordline
$y'_{1,2,3,4,5}$	Nondimensional heights of lifting surface parameters listed in Table 2
z	Operating depth of a submerged vehicle (L)
α	Angle of attack (radians); A_1/A_4
α_e	Average volume per unit of net energy output of the energy-dependent components of a vehicle $(F^{-1}L^2)$
α_p	Average volume per unit of net power output of the power-dependent components of a vehicle $(F^{-1}L^2T)$
β	A_2/A_4
γ	A_3/A_4
γ_t	Weight density of tubing material (FL^{-3})
δ	Angle of attack used for generating thickness for a supercavitating hydrofoil (radians)
ζ	Distance from the root section of a lifting surface divided by $b/2$
Δ	Designates the amount of change in some parameter
λ	Sweepback angle of a hydrofoil (radians)
μ	Poisson's ratio
ν	Kinematic viscoisity of the fluid (L^2T-1)

ρ Mass density of the fluid $(FL^{-4}T^2)$

ρ_b Average mass density of the buoyancy source $(FL^{-4}T^2)$

ρ_h Net mass density of a hydrofoil $(FL^{-4}T^2)$

ρ_o Average mass density of vehicle components which are independent of speed, range, and buoyancy requirements $(FL^{-4}T^2)$

ρ_p Average mass density of power-dependent vehicle components $(FL^{-4}T^2)$

ρ_s Average mass density of the structural material $(FL^{-4}T^2)$

ρ_v Average mass density of a vehicle $(FL^{-4}T^2)$

σ Cavitation number $= \dfrac{P-P_v}{\frac{1}{2}\rho U^2}$

σ_{cr} Incipient cavitation number, i.e., value of σ when cavitation is about to begin as σ reduces

σ_o Represents σ when $C_L = 0$, $\sigma_o = \sigma - \frac{1}{2}C_L$

$\sigma_{\lambda cr}$ Incipient cavitation number of a sweptback hydrofoil based upon the velocity normal to the span

τ Designates amount of parabolic thickness added to a hydrofoil; planform taper ratio of a lifting surface = ratio of tip chordlength to root chordlength

ϕ Thickness-to-chord taper ratio $= (t/c)_{tip}/(t/c)_{root}$

ω_α Torsional natural frequency of a submerged hydrofoil (T^{-1})

SUBSCRIPTS

x   Improved vehicle characteristic

o   Root section of a lifting surface

OTHER

$\doteq$	Approximately equal
$\lesseqgtr$	Less than, greater than, equal to, or some combination thereof
—	Terms with a bar over them are defined in Equation 307, except for $\bar{t}$

LIST OF DEFINITIONS

1. __General design objective__. A brief statement which describes the general purpose of the design.

2. __Design form__. The geometric shape of a design and all of its components.

3. __Mission specifications__. The set of all independent nondimensional requirements which must be satisfied by any design form solution to a given nondimensional design problem. (Page 11)

4. __Optimization criterion__. A nondimensional criterion whose value is to be optimized by any design form solution to a given nondimensional design problem. (Page 12)

5. __Design mission__. A nondimensional design problem which consists of a general design objective, a set of mission specifications, and an optimization criterion. (Page 11)

6. __Parameter__. A nondimensional quantity which is free to vary.

7. __Mission parameters__. The subset of mission specifications which are free to vary and which determine a set of design missions. (Page 15)

8. __Mission criteria__. The subset of mission specifications which are fixed in a set of design missions. (Page 15)

9. __Design form specifications__. The set of all independent nondimensional quantities which are required to describe

a given design form. (Page 17)

10. <u>Design form parameters</u>. The subset of design form specifications which are free to vary and which determine a set of related design forms. (Page 17)

11. <u>Design form criteria</u>. The subset of design form specifications which are fixed in a set of related design forms. (Page 17)

12. <u>Family of design forms</u>. A set of related design forms which are described by a set of design form criteria and a set of design form parameters. (Page 17)

13. <u>Design space</u>. A multidimensional Cartesian space whose coordinates are a set of independent design parameters.

14. <u>Generalized design mission</u>. A well-posed set of design missions defined by a general design objective, a set of mission parameters, a set of mission criteria, and an optimization criterion. A set of design missions is "well posed" if at least one nonempty subset is solvable by a finite number of design form families. (Page 15)

15. <u>Mission space</u>. A multidimensional Cartesian space whose coordinates are the mission parameters of a generalized design mission. (Page 16)

16. <u>Mapping criteria</u>. The set of mission criteria and the optimization criterion.

17. <u>Mapping</u>. The process of associating with a point in mission space a design form which satisfies the mapping criteria and provides the maximum degree of optimization. (Page 18)

18. <u>Region of mission space</u>. The subspace of mission space which maps into the subspace of design space which represents a specific family of design forms. (Page 20)

19. <u>Mapping relations</u>. The set of functions which are used to map from a given region of mission space into the corresponding subspace of design space. (Page 26)

20. <u>Design equations</u>. The subset of the mapping relations which is not derived directly from the optimization criterion. (Page 35)

21. <u>Overlap of regions</u>. The subspace of mission space which is shared by two or more regions. (Page 21)

22. <u>Scaling</u>. The process of changing the size of a design without changing its form or the associated design mission. (Page 83)

23. <u>Generalized scaling</u>. The process of scaling in which a specified geometric distortion is permitted and a specified change in the design mission is permitted. (Page 88)

24. <u>Optimized scaling</u>. A type of generalized scaling in which the form of a given optimum design changes as its size changes in such a manner that the resulting design is still an optimum. (Page 89)

CHAPTER I

INTRODUCTION

Nature of the Study

This study consists of the development and illustration of a systematic design procedure which can be used to solve a set of engineering design problems. The design procedure is based on a nondimensional approach. When applied to a given design field, the procedure aids in determining the diversity of possible design form families, the variations of design form within a family, the relationship between design form and the design objective, the scaling of design forms, and the general nature of research studies or new ideas which might lead to new design forms.

The design procedure is primarily oriented toward solving a set of design problems rather than solving a specific design problem. Consequently, it is most useful for determining solutions to a variety of design problems in a given field, or for determining the diversity and use of possible design forms in a given field.

The proposed design procedure should not be considered the only or the best approach to design, but merely as one approach which will hopefully be of use in the further development of design theory. The procedure may be viewed as a tool in solving a set of design problems in much the same way that a mathematical method is used as a tool in solving an equation. Considerable knowledge and ingenuity are still required in order to use the method most effectively.

Goal of Engineering Design

The objective of engineering design is to satisfy a given need with the best possible design solution. The meaning of "best" depends upon the need, and is essentially a criterion which is to be optimized such as one or a combination of the following: lowest cost, highest reliability, maximum efficiency, minimum size, longest life, minimum maintenance, etc. The given need is assumed to contain all of the information required to specify the design problem. Therefore, a design problem should specify the general nature of the problem, the performance desired (such as speed, range, flow rate, etc.), the various required operating situations (such as the operating fluid, operating time periods or cycles, characteristics of the operating environment, unsteady forces, etc.), all special criteria which the design must satisfy (such as restrictions on material or fabrication technique, requirement to match other parts, specification for minimum allowable reliability, etc.), and an optimization criterion. Summarizing, the goal of engineering design is to satisfy a given design problem which consists of a brief statement of the purpose of the design, and the specification of the performance objectives, the required operating situations, all special design criteria, and an optimization criterion.

Background

Considering the many methods of engineering design presented in the various design handbooks, reference books, and reports in specialized fields, surprisingly little information relates to engineering design procedures which are based on a nondimensional analytical[1] approach and which apply to a wide variety of design fields.

Zwicky's morphological method (2) is a noteworthy semi-analytical approach wherein a general design solution is first established which consists of basic design components such as the power source, structural material, sensing systems, propulsor, etc. All possible types of each component are then listed, and finally, each resulting combination of types of components is analyzed as a possible design solution to the design problem.

---

[1] An analytical approach, as used here, is meant to exclude design approaches based principally upon random search techniques, some examples of which are the Monte Carlo method, game theory, or any of the related methods used in systems analysis. However, random search techniques can be useful in certain portions of an analytical approach and have been found useful as the basis for design problems which are well understood. One very useful type of random search technique was developed by Mandel (1) which, although applied to ship design, could be readily generalized for use in other design fields. This approach consists of solving a specific design problem by utilizing a digital computer to consider a large number of designs consisting of variations of a predetermined set of design variables. The performance of each design is determined by a special computer program, and the design which provides optimum performance is selected.

Some fundamental aspects of the design process are presented by McLean[1] (3), who believes that simple and reliable design is an art which requires a designer with creative talents who understands his field, and has the freedom, time, and encouragement to express his talents. It is because of the many reasons presented in (3) that any systematic procedure of design will still require considerable knowledge and ingenuity by the designer in order to be most effective. Some of the design concepts presented by McLean (3) are: (a) a broad statement of the problem will leave the designer much more freedom in creating novel solutions and reduce the chance of being channeled into a specific type of solution; (b) the designer should gain a thorough understanding of the factors which set the limits on the design problem; such factors may be natural physical limits or limits imposed by the state of the art; (c) the designer must develop an understanding of the trade-offs which may exist under an overall limit where an improvement in one desirable characteristic leads to a decreased ability to fulfill another desirable characteristic; and (d) a simple design is anything but simple in its creation, and often appears to be so simple and logical that it is difficult to imagine why so much time and effort were required for its development.

A possible approach to a generalized design method is presented in a paper by Gabrielli and Von Kármán, called "What Price Speed?" (4). This paper contains a graph of empirical data showing specific power

_____

[1] Inventor of the well-known Sidewinder air-to-air missile.

(horsepower per ton) as a function of maximum speed for a wide variety of land, sea, and air vehicles. A limiting relationship between minimum specific power and maximum speed was found which indicates the minimum price in power that must be paid for increased speed, independent of the vehicle type. The graph also shows the type of known vehicle which requires the minimum specific power to achieve a given maximum speed. This result suggests that non-empirical methods might be developed which would provide similar and other kinds of general limiting relationships.

Davidson (5) performed a generalized theoretical analysis of ships, which also included a brief study of airplanes, in which he showed that neither ships nor airplanes were necessarily restricted by the empirical limiting line of specific power versus speed presented in (4). In agreement with Gabrielli and Von Kármán, Davidson showed that the limiting line is determined by maximum size, which in turn is primarily structurally limited. Significantly, his analysis showed that the limiting line could be exceeded by large ships and airplanes which lie within the scope of current technology and would not require major improvements in materials. His general-ized analysis is of interest because he utilizes the nondimensional method to considerable advantage and establishes the beginning of an approach for analyzing a variety of design forms.

A more systematic nondimensional approach to design is described in a recent report by Wislicenus (6). The specified operating conditions (i.e., design requirements) of a design problem are placed in nondimensional form and, through the use of physical

relationships, are equated to functions of nondimensional design
variables. A solution of the resulting set of equations helps to
provide the desired design form. The approach clearly demonstrates
the significance of a nondimensional method. The nondimensional
operating conditions are essentially similarity relationships which
permit broad scaling of the resulting design forms. Some of the
design concepts presented by Wislicenus (6) are: (a) form design
should not be pursued primarily on an intuitive basis but rather on
a rational approach as far as possible; (b) designers should recog-
nize the existence of related design forms in technology and nature[1];
(c) geometrically similar design forms may be classed together
regardless of size; (d) a large number of systematically related
design forms exist which may be called families and which are
characterized by variations of a dimensionless form or performance
parameter; (e) the specified dimensionless operating conditions must
be related in some rational fashion to dimensionless elements of the
design form; (f) the quantity of design variables cannot be expected
to completely describe a design form, but need not be large for
preliminary design purposes; (g) in the field of design, any
conceivable class of objects can be defined not only by its physical
characteristics but also by a set of operating or design requirements; and (h) there is, and must be, a relation between the design

---

[1] Such forms in nature are acknowledged in (6) to nave been
suggested by John Erwin of the General Electric Company,
Cincinnati, Ohio. A very complete nondimensional analysis
of related forms in nature has been conducted by
W. R. Stahl (7).

requirements and the physical characteristics of the object to be designed; otherwise, a design problem would have no solution.

Further use of the similarity relationships derived by Wislicenus was made by Werner (8) in a report on the analysis of airplane design. Aircraft form characteristics and families of aircraft types were related to specific values of various similarity parameters. This analysis permitted many conclusions to be drawn regarding trends in aircraft design form as a function of weight, speed, engine type, and other variables.[1]

Other significant approaches to design were suggested by D. P. Hoult[2] during informal discussions. He proposed that a multidimensional space be set up (consisting of nondimensional variables which affect design form) and that regions in this space be found which correspond to certain families or groups of design form. Also, he suggested a way in which group theory might be utilized in design.

Statement of the Problem

The objectives of this study were to:

1. Establish the basic concepts of a general and complete

---

[1] The results of the analysis showed that none of the modern very large airplanes yet exceed the limit line established by Gabrielli and Von Kármán.

[2] Associate Professor in the Department of Aerospace Engineering, The Pennsylvania State University, University Park, Pennsylvania. Currently Associate Professor in the Department of Mechanical Engineering, Massachusetts Institute of Technology, Cambridge, Massachusetts.

engineering design theory based on an analytical nondimensional approach.[1]

2. Illustrate the theory by a number of different but general design examples.

3. Develop methods permitting the establishment of areas in which inventions or research studies are still needed.

4. Establish the basic concepts of the design theory in a rigorous manner thereby permitting mathematical treatment of design problems as far as possible.

Outline of the Text

The design procedure is developed in Chapter II, and is outlined in Chapter III which also contains a list of its advantages and brief illustrations showing how the procedure is used in solving simple design problems. Chapter IV contains a more complete discussion of the design procedure, its use, related topics, and comments on the future development of design theory. Three relatively complex design examples are presented in Chapter V, Appendix A, and Appendix B which treat the design of submerged bodies, airplane wings and hydrofoils, and hydrofoil cross sections, respectively. The latter design problem is included because it may be considered partially a research problem and therefore demonstrates one of the

---

[1] The previously mentioned nondimensional design theories based on an analytical approach lack completeness because a specific design procedure is missing and the optimization criterion is not included.

ways in which the design procedure is applicable to research. The results of the study are summarized in Chapter VI which also contains the conclusions.

CHAPTER II

DEVELOPMENT OF THE DESIGN PROCEDURE

This chapter presents the principal concepts on which the design procedure is based. The nondimensional approach to design is discussed, and the notion of a generalized design mission is presented. Any questions regarding the design theory or procedure which are not clearly answered in this chapter may be found answered in Chapter III or Chapter IV.

Design Problem

As mentioned in Chapter I, a design problem typically consists of: (a) a brief statement which describes the general purpose of the design (i.e., a general design objective), (b) a set of independent[1] requirements which must be satisfied by any design solution (i.e., design problem specifications), and (c) a criterion which is to be optimized by any design solution (i.e., an optimization criterion). The independent requirements consist of certain desired performance characteristics, prescribed operating situations, and prescribed design characeristics.

---

[1] A quantity which belongs to a set of quantities is said to be independent if its value cannot be calculated from the values of the other quantities belonging to the set.

Designers may be presented with a great variety of design problems. Some problems may be so specific that little is left to design, while other problems may be so broad that the designer must consider a wide variety of design forms before selecting the form which best satisfies the design goal. The design procedure developed herein can be applied to either specific or to broad design problems.

Design Mission

A design mission is defined as a nondimensional design problem which consists of a general design objective, a set of independent mission specifications (i.e., a set of nondimensional design problem specifications), and an optimization criterion (which is assumed to be nondimensional from here on).

Mission specifications. By nondimensionalizing the design problem specifications, a considerable number of design problems can be collapsed into relatively few design missions. Instead of specifying the dimensional values for each of the design problem specifications, corresponding values for nondimensional groupings of them are specified. Examples of such nondimensional groupings, or mission specifications, are pump specific speed, airfoil lift coefficient, vehicle Reynolds number, ship Froude number, hydrofoil cavitation number, airplane Mach number, etc. Consider Mach number, for example, and note the large number of combinations of airplane speed and local speed of sound which are required to replace a single value of the Mach number (which represents the influence of compressibility). Similarly, consider all of the values of velocity,

length, and kinematic viscosity which must be used to replace a
given Reynolds number (which represents the influence of fluid
viscosity in a dynamic problem). Consequently, it is seen that the
use of a nondimensional parameter not only permits information to
be condensed, but may also be used to represent the influence of a
certain physical phenomenon. Furthermore, the value of a nondimensional parameter gives true meaning to such terms as "high speed",
"low viscosity", "large size", etc. An example of the use of the
nondimensional approach in solving design problems related to
various types of turbomachinery is presented by Wislicenus (9).

Design forms. Some of the mission specifications may
consist of specified design characteristics. Since the mission
specifications are nondimensional, all specified design characteristics must be nondimensional. Therefore, the specified design
characteristics must describe some aspect of the design form,
where "design form" is defined as the geometric shape of a design
and all of its components. When speaking of design form, size is
no longer significant. Examples of design form characteristics are
wing aspect ratio or sweep angle, ship length to beam ratio, hydrofoil thickness-to-chord ratio, number of teeth in a gear, etc.
Two designs are said to be geometrically similar if their forms are
equal.

Optimization criterion. The optimization criterion is a
single nondimensional criterion which is a function of one or more
nondimensional independent optimization parameters, such as maximun
efficiency, minimum operating cost ratio, etc. The optimization

criterion is evaluated for each of the many possible design form solutions. A design form solution is defined as a design form which provides the optimum value of the optimization criterion. In the few cases where two or more design forms are found to provide the same optimum value for the optimization criterion, they are to be considered equally valid solutions.

The optimization criterion must reflect the relative importance of the various independent optimization parameters. For example, assume that the nondimensional construction cost and nondimensional operating cost of a design are to be minimized, and the efficiency is to be maximized. Since the parameters are independent, it is impossible, in the general case, to find a single design form which optimizes each of all three parameters. Consequently, the designer must determine the relative importance of each parameter. For example, suppose he decides that linear relationships can be utilized and that plus 5% in efficiency is equivalent to minus 0.18 in the construction cost parameter, or to minus 1.38 in the operating cost parameter. In this case the value of the optimization criterion Q would be:

$$\text{Value of } Q = \frac{\% \text{ efficiency}}{5.0} - \frac{\text{construction cost}}{0.18} - \frac{\text{operating cost}}{1.38}$$

The best design form would be the form having the highest value of Q. This approach to evaluating Q can easily be modified to include variable weighting factors. For example, if plus 5% in efficiency when the efficiency is 80% is considered equivalent to plus 10% in efficiency when the efficiency is 50%, the weighting factor for efficiency is a function f_η of the efficiency; consequently, the efficiency term in Q would be written as: (efficiency) $\div$ (f_η of the efficiency). The same approach is applicable to the other terms in Q.

The optimization criterion is not always best expressed as a simple sum. In evaluating the effectiveness of a missile, Smith and King (10) developed an expression called mission success, abbreviated as M. S., which is defined as:

$$M.~S. = R \times A \times P$$

where $R = e^{-t/\Theta}$, $A = \frac{\Theta}{\Theta + \phi}$, P = kill probability against specific threat environments, t = mission time, Θ = mean time between failures, and ϕ = mean time to repair. Consequently, if several missile designs are being considered, and no other optimization parameters are to be considered, the best design is defined as the one with the highest value of M. S.

The optimization criterion may be a function of nondimensional performance characteristics, operating situation variables, or design characteristics. Efficiency and kill probability are examples of performance characteristics, mission time is an operating situation variable, and construction cost or mean time to repair are examples of design characteristics. Note that all quantities are functions of the design.

Subdesign missions. A specific design mission can (and often should) be separated into subdesign missions. For example, an airplane design mission can be separated into subdesign missions such as the design of the wing, landing gear, motor, etc. When separating a design mission into subdesign missions, the designer must be careful to account for all significant interactions between the various subdesign missions. For example, the design of an airplane wing cannot be completely isolated from the rest of the airplane since it may be required to house the motor, fuel, and landing gear; however, its design can still be isolated in certain respects and treated as a separate design mission with certain restrictions (mission specifications) imposed upon it. By correctly separating a complex design mission into subdesign missions, the complexity is often reduced considerably. Therefore, the designer should always consider the possibility of establishing subdesign missions.

Generalized Design Mission

A given design mission can be generalized into a set of design missions by permitting some or all of the mission specifications to vary. Such a set of design missions is called a generalized design mission if at least one mission of the set can be satisfied by a finite number of design form solutions. A generalized design mission may have only one mission criterion which varies, or it may have more, depending upon how general the designer wants to make it. The mission specifications which vary are called mission parameters, and the specifications which remain fixed are called mission criteria. If all of the mission specifications remain fixed, the design mission is not generalized. Alternatively, if all of the mission specifications are changed into mission parameters, the given design mission is completely generalized. Note that when setting up a generalized design mission, the designer does not need to begin by considering a given design mission. The given design mission was introduced to help explain the mechanism of setting up a generalized design mission; however, beginning with a typical design mission is often a useful approach in setting up a generalized design mission.

A generalized design mission is therefore seen to consist of a general design objective, a set of independent (variable) mission parameters M, a set of independent (fixed) mission criteria C, and an optimization criterion Q. All parameters and criteria are nondimensional. The generalized design mission is solved by associating a design form (or forms) which best satisfies the optimization criterion, with each of the many design missions comprising the generalized design mission.

Mission Space

A specific design mission results when a specific value is assigned to each of the independent mission parameters belonging to a generalized design mission. If not all mission parameters are fixed, a set of design missions results which is a subset of the set of design missions comprising the generalized design missions.

The set of mission parameters in a generalized design mission may be considered to be the coordinates of a multidimensional Cartesian space called mission space. A fixed value of each mission parameter is a point in mission space, and represents (together with

the mission criteria, an optimization criterion, and a general
design objective) a specific design mission, as discussed above.
Similarly, a subspace of mission space represents a set of design
missions; such a set is a subset of the set of design missions
represented by the entire mission space. The concept of mission
space is introduced because it aids in visualizing a multidimensional
set of design missions, and is shown later to permit each of the
many design solutions to be clearly associated with the design missions
by means of graphs which represent sections or subspaces of mission
space.

Design Space

As stated earlier, a design form can be described by a set
of design form specifications. These specifications should not be
so detailed that every bolt and rivet in the design is described,
but should carry the basic information necessary to permit a typical,
experienced designer to complete the design in detail. In other
words, the design form specifications should be sufficiently complete
to describe a preliminary design.

<u>Family of design forms</u>. If some of the set of design form
specifications which describe a given design form are permitted to
vary, a family of design forms results. The design form specifications which vary are defined as design form parameters, and the
design form specifications which remain fixed are defined as design
form criteria. Therefore, a family of design forms may be defined
as a set of related design forms which are described by a set of

(fixed) design form criteria and a set of (variable) design form parameters. A specific design form results when a specific value is assigned to each design parameter.

Definition of design space. The set of all design form solutions to a given generalized design mission can be described by a set of (fixed) design form criteria and a set of (variable) design form parameters. The design form parameters are considered to be the coordinates of a multidimensional Cartesian space called design space. A fixed value of each design parameter is a point in design space, and represents a single design form. A family of design forms is represented by a subspace of design space.

Mapping

Mapping is defined as the process of associating with a point in mission space an optimum design form which satisfies the mapping criteria. The mapping criteria consist of the mission criteria and the optimization criterion. By "optimum" is meant a design form which provides the optimum value of the optimization criterion. Therefore, mapping is merely the process of associating a design form with a given point in mission space, where the associated design form is defined as an optimum design form solution. The mapping process is schematically illustrated in Figure 1. The mission space is represented by the mission parameters m_1 through m_p and the design space is represented by the design parameters d_1 through d_q. The number of coordinates in the two spaces is not necessarily equal.

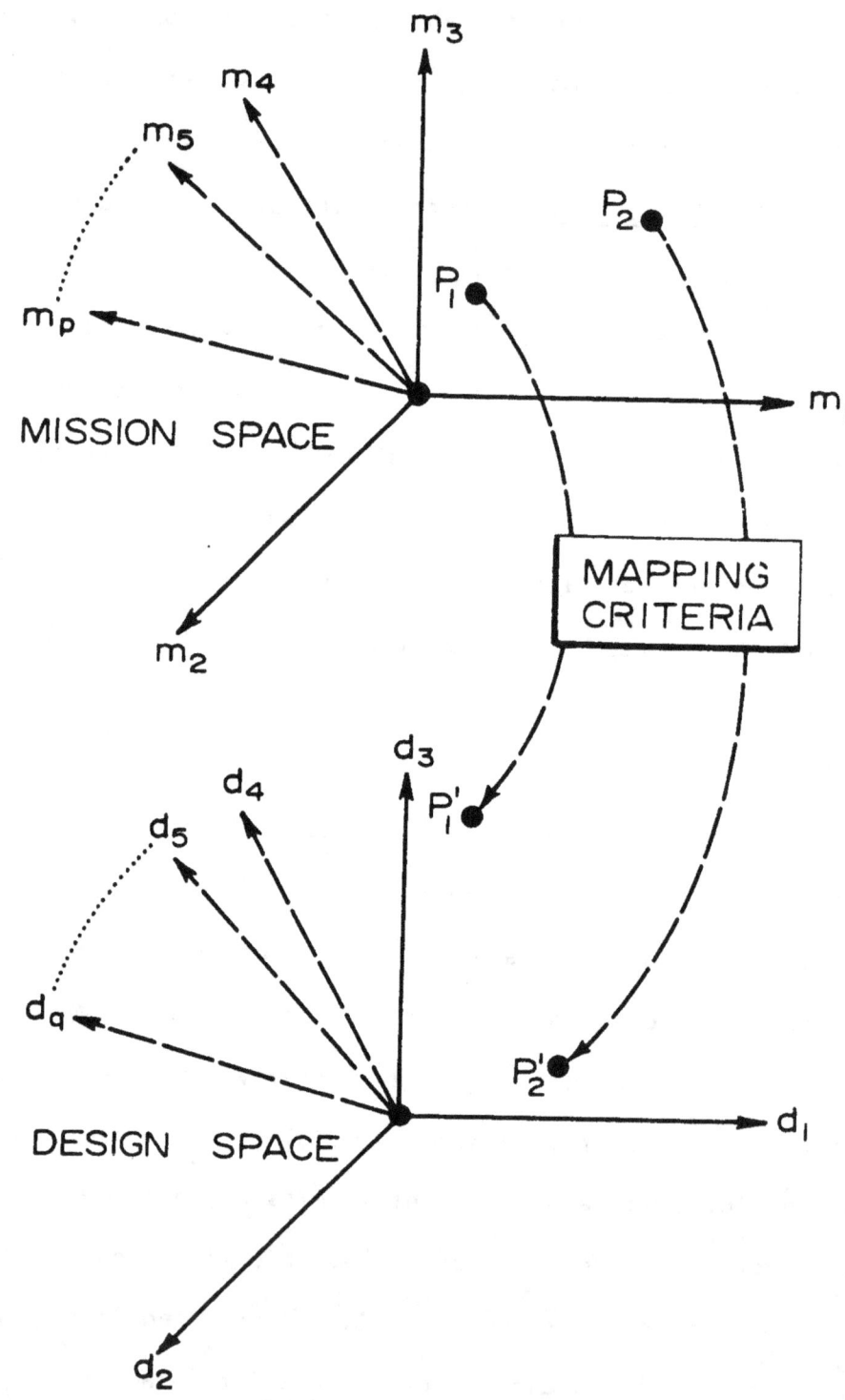

Figure 1 - Schematic illustration of the mapping process

Regions of Mission Space

Assume that each point in a given subspace of mission space maps into a single corresponding design form. Then, a specific design form will be associated with a selected point in the mission subspace. Now consider a nearby point in mission space. This nearby point will map into either the same design form, a slightly modified design form, or an entirely different design form. Similar mappings result from other nearby points. The portion of the given subspace which maps into a single family of related design forms is called a region of mission space. All points lying outside of a given region in the assumed subspace will lie in a different region and therefore map into a different family of design forms.

Now consider a mission space which corresponds to an arbitrary generalized design mission. Such a mission space will be found to contain one or more regions, where a region is defined as the subspace of mission space which maps into a single family of design forms.[1]

Void regions. It is entirely possible that regions of mission space will be found whose points cannot be mapped into any known design form or any form which the designer might invent. Such regions are called void regions, and indicate that either further invention or research is needed, or that no possible solution exists.

---

[1] According to this definition, any subspace where two or more distinct design forms correspond to each point is considered to consist of two or more regions which share that subspace. Such regions may or may not be equal in size, shape, and location since they may extend beyond that subspace.

Overlapping regions. Regions of mission space may be found which overlap. The subspace where such an overlap occurs is the locus of all points to which more than one design form corresponds. The number of different design forms which correspond to each point in mission space where regions overlap is equal to the number of overlapping regions because one distinct design form corresponds to each region.

Typical Design Forms

The selection of typical design forms which correspond to various regions of mission space is an important step because knowledge of such typical design forms is required before a given region can be mapped. The typical form corresponding to a given region of mission space is dependent upon the state of the art in most cases; consequently, the selection of a typical design form depends upon the designer's knowledge of the design field, and in some cases, upon his inventive ability to improve upon known typical forms or to develop new typical forms when none are known.

A comprehensive literature search and contact with specialists in the specific design field may serve to provide the necessary knowledge. Systematic methods for inventing typical design forms are lacking. One approach, however, that may be of help is to first study a variety of arbitrary forms using physical relationships, logic, and intuition to determine if any are acceptable. The forms of the more acceptable candidates are then varied, and the process is continued until a reasonably good set of typical design forms emerge. The use of Zwicky's morphological method (2) mentioned earlier may be of some help in the more complex problems. Keeping in mind the objectives of simplicity and reliability discussed by McLean (3) should help in selecting or modifying possible design forms. In some cases, the use of a computer for the random selection and evaluation of possible design forms may be worth considering; the use of knowledge and intuition in the programming may save considerable time.

Region Boundaries

In some cases, the boundary between two adjacent regions will map into the same subfamily of design forms. When this occurs, the design forms corresponding to mission space will vary smoothly and continuously as a function of position in mission space, and the boundary line between two regions in mission space will be sharply defined. On the other hand, the boundary in mission space may correspond to the crossover from one design form in one family to a distinctly different design form in another family. Such a boundary will be sharp if it represents the limiting design form of a family which is significantly more acceptable than the other family. The boundary under consideration will generally be hazy, however, if the two design families corresponding to adjacent regions become equally acceptable at the boundary. The reason is that their relative acceptability may change very slowly in the region near the boundary, thereby making the boundary difficult to locate precisely. Also, as indicated earlier, two design families may be exactly equally acceptable and therefore correspond to the same subspace of mission space, in which case the boundaries would overlap.

Critical values of mission parameters. An aid in locating regions in mission space is to first determine which of the mission parameters have critical values. By "critical value" is meant a value above which the corresponding design form is designed by one set of rules, and below which it is designed by an entirely different set of rules. These critical values could either represent a natural physical limit which will not change with time, or they may result from a man-made limit which may vary with time. In either case, the critical values are treated the same way.

Examples of parameters which generally have at least one critical value are cavitation number, Reynolds number, and production rate. The cavitation number in a hydrofoil design problem has a critical value since one set of design rules is used at zero cavitation number, and an entirely different set of design rules is used at high cavitation numbers. Similarly, Reynolds number contains a critical value, since below a certain Reynolds number the boundary layer over the front of a streamlined body moving through a fluid is laminar, while above a certain Reynolds number the boundary layer is largely turbulent. The optimum form of the body is significantly different in the two cases since different design rules are followed. In a typical economic problem, the production rate normally has at least two critical values. The lower value would correspond to the point below which no design form is economically feasible to produce.

The upper critical value might correspond to a higher production rate where a different structural material could be utilized, permitting a change in design form which may result in better performance. Alternatively, the upper critical value could represent the crossover from hand production to machine production which might permit major design changes to occur.

Location of boundaries in mission space. The location of boundaries in mission space can often be accomplished without developing a complete mapping relationship between mission space and design space. After determining the typical design form and the basic phenomena which cause a boundary to appear in mission space, it is often possible to utilize physical relationships to determine the exact location of the boundary and the design forms corresponding to the boundary.

One possible aid in determining design form families and the corresponding region boundary in mission space is to use the fact that two design families which might correspond to opposite regions adjacent to a boundary often merge into the same design subfamily along the boundary. Another aid in locating a boundary is to determine if the boundary corresponds to the limiting form of one specific design family; if so, the boundary can be located by investigating only that one family. The designer should also be aware of void regions since the boundaries to such regions are sometimes easy to locate, and should be located early in the design process.

Figure 2 illustrates how a boundary in a three-dimensional

25

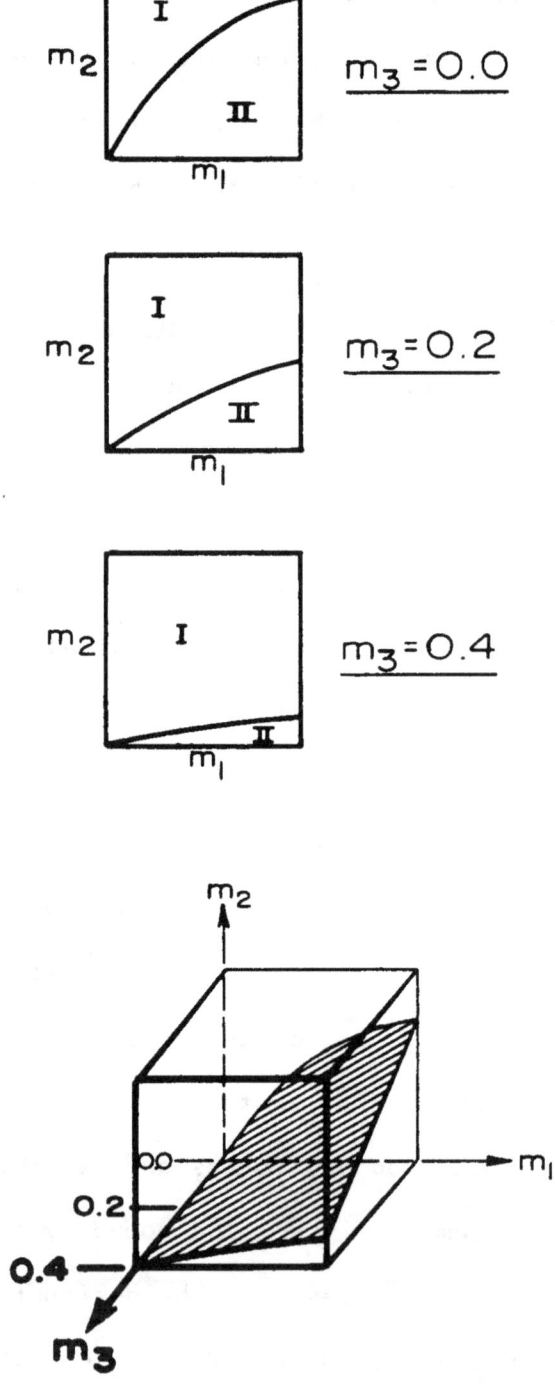

Figure 2 — Illustration of a boundary in three-dimensional mission space

subspace of mission space can be considered either as lines in a series of two-dimensional sections or as a surface in three-dimensional space. Chapter III contains design examples which may help clarify the procedure of locating boundaries.

Mapping from Mission Space to Design Space

Once a certain design form family has been established in a general manner, and a corresponding region in mission space has been determined, the designer can proceed to map the region into specific design forms. To do this, physical relationships must be established between the mission parameters and the design parameters.

Mapping relations. If a set of relations can be developed for mapping a certain region of mission space, the relations (called mapping relations) can be used to map any of the infinite number of points in that region. Unfortunately, the mapping relations become more difficult to develop as the number of dimensions in the region considered is increased. Consequently, if a set of mapping relations cannot be developed for an entire region, the designer can begin by developing mapping relations for subspaces within a region and then try to develop a more complete set of mapping relations by considering other subspaces within the region.

Mapping sequence. It has been found helpful to begin mapping mission space by proceeding from the simple and most familiar design missions to the complex and least familiar design missions. If consecutive mappings are adjacent, the results of a previous mapping can be used as a boundary mapping for the new mapping; this procedure is often of great help in determining the mapping relations for unfamiliar subspaces of mission space. An excellent place to begin mapping is the simplest point in the entire mission space. This point is generally the one whose coordinates are such that the corresponding mission parameters do not affect the design form; its coordinates are usually zero or infinity. The next simplest subspaces of mission space to map are other simple points and some of the coordinate axes. Following that, the three coordinate planes of a three-dimensional subspace might be mapped. Then, by selecting planes parallel to one of the coordinate planes, most of an entire three-dimensional subspace can be mapped. By similarly studying other three-dimensional subspaces in a given region, valuable information on the entire design picture of a multidimensional region can be obtained.

The coordinates which are generally most important to include in the various mappings of mission space are those parameters which represent the phenomena that most strongly affect the design form. Examples are parameters which include the design stress, speed, applied forces, etc. Also, considerable time can be saved in mapping by first finding those parameters which have ranges of values that do not appreciably affect the design form. Only one typical value of the parameter in each of the uncritical ranges must be investigated.

Illustration of a mapping. The best way to illustrate a mapping appears to be a graphical presentation. For example, if a particular line in mission space has been mapped, a two-dimensional graph can be drawn where the abscissa represents the particular line and the ordinate provides the value for various curves on the graph where each curve represents a particular design parameter or the optimization criterion of the corresponding design form solutions. Similarly, the mapping of a plane in mission space can be illustrated by a graph (or graphs) showing one mission parameter plotted against the other with a series of lines superimposed on the graphs showing the values of the associated design form characteristics and the optimization criterion. Three-dimensional mappings can sometimes be illustrated by a single two-dimensional graph, always by a series of two-dimensional graphs, and sometimes by a three-dimensional graph or a drawing of a three-dimensional graph. Figure 3 illustrates a typical two-dimensional mapping where ficticious values are plotted for the optimization criterion Q and one design parameter d_1.

Dimensionalizing the Design

So far, the design mission has been treated in nondimensional form. The final objective, in general, is a dimensional, full scale solution. In a typical design problem, the values of all specifications are dimensional. Consequently, all the designer must do to solve a specific design problem, assuming that the generalized design mission has been solved, is to: (a) calculate the mission parameters which correspond to the given design problem, (b) find the resulting values for the design form parameters from the solution to the generalized design mission, and (c) calculate the desired dimensional design characteristics by substituting the specified dimensional quantities into the design form parameters.

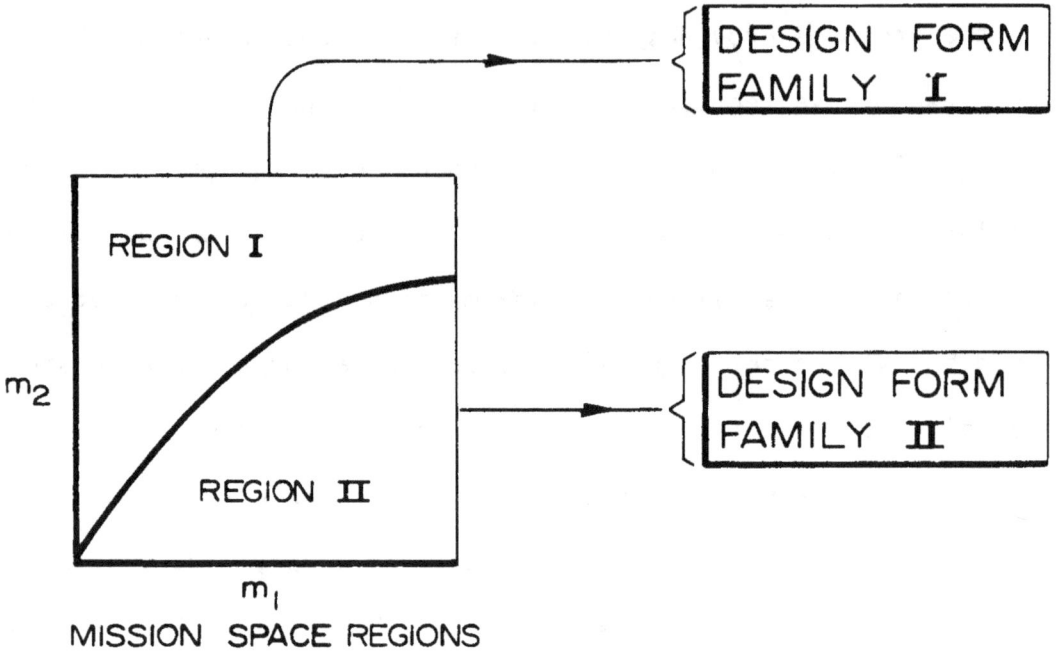

MISSION SPACE REGIONS

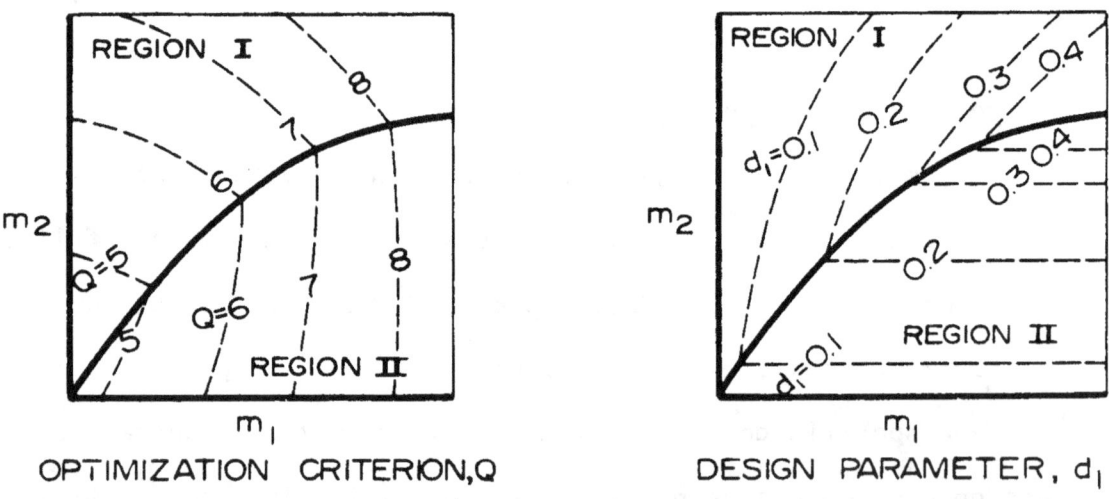

OPTIMIZATION CRITERION,Q

DESIGN PARAMETER, d₁

Figure 3 - Illustration of a two-dimensional mapping from a
mission space consisting of two regions

Classification

One of the advantages of this design method is that the different mission parameters, design form parameters, and regions of mission space which result from solving a generalized design mission can be used to classify operating situations, designs, and design families into natural classes which have physical meaning. Three simple design examples are presented in Chapter III which illustrate the classification concept.

Mathematical Representation

The design procedure is now placed on a more formal basis. Let the set of all significant (nondimensional) mission parameters m_1, m_2, ..., m_p be represented by $M = \{m_i\}$ where $i = 1, 2, ..., p$. Similarly, let the set of all significant (nondimensional) design form parameters d_1, d_2, ..., d_q be represented by $D = \{d_j\}$ where $j = 1, 2, ..., q$. Also, let the set of all significant (non-dimensional) mission criteria c_1, c_2, ..., c_r be represented by $C = \{c_k\}$ where $k = 1, 2, ..., r$. Finally, let the set of all significant (nondimensional) optimization parameters q_1, q_2, ..., q_s be represented by $\{q_\ell\}$ where $\ell = 1, 2, ..., s$. The optimization criterion Q is expressed as a function of the q_ℓ. Each of the sets $\{m_i\}$, $\{d_j\}$, $\{c_k\}$, and $\{q_\ell\}$ is assumed to be composed of independent parameters. Significant d_j are all design parameters which have a definite relationship with the m_i, c_k, and q_ℓ.

As stated earlier, M is considered to be a multidimensional Cartesian space in which the coordinates are the mission parameters m_i.

A point in such a space represents a specific design mission (when combined with a general design objective, the set C, and Q), and is determined by an ordered sequence of values of the m_i.

Similarly, D is considered to be another multidimensional Cartesian space in which the coordinates are an independent set of the d_j. A point in this space represents a specific design form.

The design process is looked upon as a mapping from the M space (mission space) into the D space (design space). The optimization criterion Q and the mission criteria C are the mapping criteria. Since the design form is a function of the mission parameters, mission criteria, and the optimization criterion, the relationship can be written as

$$D = D \ (M, \ C, \ Q) \tag{1}$$

Physical relationships. Assume that a typical design form has been found and the corresponding region in mission space has been established. The method introduced here for relating design forms to mission parameters is a modification of the method proposed by Wislicenus (6). Considerable knowledge of the relevant physical phenomena is required in order to map points from mission space into design space. Such knowledge results in the following relationships:

$$m_1 \gtreqless f_1 \ (d_1, \ d_2, \ \ldots, \ d_q)$$

$$m_2 \gtreqless f_2 \ (d_1, \ d_2, \ \ldots, \ d_q) \tag{2}$$

$$\vdots$$

$$m_t \gtreqless f_t \ (d_1, \ d_2, \ \ldots, \ d_q)$$

$$c_1 \gtreqless g_1 (d_1, d_2, \ldots, d_q)$$

$$c_2 \gtreqless g_2 (d_1, d_2, \ldots, d_q) \tag{3}$$
$$\vdots$$
$$c_r \gtreqless g_r (d_1, d_2, \ldots, d_q)$$

$$Q = Q (q_1, q_2, \ldots, q_s) \tag{4}$$

$$q_1 = q_1 (d_1, d_2, \ldots, d_q)$$

$$q_2 = q_2 (d_1, d_2, \ldots, d_q) \tag{5}$$
$$\vdots$$
$$q_s = q_s (d_1, d_2, \ldots, d_q)$$

where $m_1, m_2, \ldots, m_p$ are the mission parameters; $d_1, d_2, \ldots, d_q$ are the design form parameters; $c_1, c_2, \ldots, c_r$ are the mission criteria; $q_1, q_2, \ldots, q_s$ are the optimization parameters; p and q are the number of dimensions of the mission space and design space, respectively (which are not necessarily equal); r and s are the number of fixed criteria and optimization parameters, respectively; $\gtreqless$ is a symbol meaning equal to, greater than, less than, or a combination thereof; f and g with subscripts represent functions; and Q, or q with a subscript, placed to the left of a sequence of symbols also represents a function.

The relationships of Equations 2 through 5 are purely symbolic, since to mathematically express each parameter as such a clear-cut function may not be possible in all cases. In other words, two parameters may be so completely interelated in two different expressions that neither could be solved as the kind of single function shown above, even though the parameters are theoretically separate and independent functions.

Relationships between the mission and design form parameters. The set of equations labeled Equation 2 associates each mission parameter with the set of design parameters. Such an association is possible since it has been assumed that the general design form is known; therefore, if sufficient knowledge exists, each mission parameter can be expressed as a function of the design form which, in turn, is described by d_1, d_2, ..., d_q.

An inequality sign will appear in Equation 2 whenever a critical value exists in a given mission parameter. The reason will become evident in the examples which are presented later in this chapter. If a mission parameter has no critical value, the symbol $\gtreqless$ will always become an equal sign.

Relationships between the mission criteria and the design parameters. Since the nature of the (fixed) mission criteria c_k is the same as that of the (variable) mission parameters m_i, the relationships between the c_k and the d_j are developed in the same way as the relationships between the m_i and the d_j.

Trivial relationships. Some of the relationships for the c_k and the fixed m_i (i.e., m_i which are fixed in a given region of mission space) may be trivial. For example, if one of the c_k or fixed m_i is a parameter consisting of the nondimensional design stress, and if the selected value of that parameter is infinity, then the structural material would be considered to be infinitely strong; consequently, any relationship between the design form and the design stress would be trivial since, from the viewpoint of stress, the design could have any form. Similarly, if the operating cavitation number is selected as infinity for a given region of mission space, cavitation could never occur; consequently, the relationship between design form and cavitation number would be trivial since, from the viewpoint of cavitation, the design could assume any form.

Void relationships. Note also that some of the fixed m_i relationships may be void since the selected value of the particular m_i may be impossible to satisfy by any known design form or any design form which the designer might invent. In this case, the selected region of mission space has no known design solution. Therefore, that particular region of mission space is said to be a void region since it is void of a solution as far as the designer is concerned. In case any of the (fixed) mission criteria c_k lead to a void relationship, the entire set of design missions is void.

Relationship between the optimization criterion and the design parameters. Now consider Equations 4 and 5. These equations show that the optimization criterion Q can be expressed as a function of the design parameters d_j. Since the various optimization parameters q_ℓ are assumed to be independent, the number of optimization parameters must be less than or equal to the number of design parameters. If this relationship is not satisfied, the optimization criterion has not been properly defined in terms of independent q_ℓ.

Design equations. Since the known relationships (i.e., all relationships resulting from the c_k and the fixed m_i) of Equations 2 through 5 result in reducing the quantity of unknown d_j and in trivial relationships (assuming that none are void), the following unknown parameters and relationships remain:

$$m_1 \gtreqless f_1 (d_1, d_2, \ldots, d_u)$$

$$m_2 \gtreqless f_2 (d_1, d_2, \ldots, d_u) \tag{6}$$
$$\vdots$$
$$m_t \gtreqless f_t (d_1, d_2, \ldots, d_u)$$

$$Q = Q (d_1, d_2, \ldots, d_u) \tag{7}$$

where u is the number of unknown d_j remaining after evaluation of the c_k and the fixed m_i, and t is the number of dimensions of the region of mission space being considered. (The m_i and d_j have been reordered to permit sequencing of the subscripts.) Calling the nontrivial relationships in Equation 6 the design equations, the quantity of the design equations is equal to, or less than, the number of dimensions of the selected region of mission space. This quantity will seldom be greater than two or three, and must always be less than or equal to the number of unknown d_j for the reason presented below.

The number of unknown d_j cannot be determined because an undetermined number of known relationships will have been trivial. However, in any given analysis, the designer can readily determine how many unknown d_j remain after the relationships have been reduced to the design equations and Q. If the number of unknown d_j is exactly equal to the number of design equations, Q is no longer needed, and the unknown d_j can be readily evaluated. If the number of unknown d_j is less than the number of design equations, the generalized design mission was not specified correctly since some of the mission parameters or mission criteria were not independent, as assumed. Finally, if the number of nontrivial design equations is fewer than the number of unknown d_j, the expression for Q must be used in order to obtain additional relations and solve the problem.

Treatment of the optimization parameter Q. The expression for Q is clearly seen to require a much different treatment than the design equations in the process of finding a design solution because it is the parameter which must be optimized. The expression for Q is generally not an equation which can be solved for one of the d_j like the design equations. However, Q can be utilized to provide all of the additional relations needed to solve the problem. For example, if there are three unknown d_j and only two design equations, Q can be utilized to provide the missing relation. Similarly, if five unknown d_j exist and only two design equations are available, Q can be utilized to provide the three additional relations needed to solve the problem. Because the design equations are defined to be not trivial, they can be solved for some of the unknown d_j and substituted into Equation 7 to obtain

$$Q = Q\ (d_1,\ d_2,\ \ldots,\ d_{u-t},\ m_1,\ m_2,\ \ldots,\ m_t) \qquad (8)$$

where the d_j have been reordered again to permit sequencing of the subscripts. Since Q is to be either maximized or minimized, the following additional equations are useful:

$$\frac{\partial Q}{\partial d_1} = 0 = Q_1\ (d_1,\ d_2,\ \ldots,\ d_{u-t},\ m_1,\ m_2,\ \ldots,\ m_t)$$

$$\frac{\partial Q}{\partial d_2} = 0 = Q_2\ (d_1,\ d_2,\ \ldots,\ d_{u-t},\ m_1,\ m_2,\ \ldots,\ m_t) \qquad (9)$$

$$\vdots$$

$$\frac{\partial Q}{\partial d_{u-t}} = 0 = Q_{u-t}\ (d_1,\ d_2,\ \ldots,\ d_{u-t},\ m_1,\ m_2,\ \ldots,\ m_t)$$

However, some of the expressions in Equation 9 may be trivial since they may not correspond to an optimum point.[1] A trivial result means that the particular d_j being considered is to be either maximized or minimized, depending upon its relationship in Equation 8. In case none of Equations 9 is trivial, the equations can be solved and the design solution obtained as a function of $m_1,\ m_2,\ \ldots,\ m_t$.

---

[1] Methods of advanced calculus can be used to determine if each of Equations 9 corresponds to an optimum value. Alternately, an inspection of the physical situation may show whether the resulting values are optimum points.

In case one or more of Equations 9 are trivial, the affected d_j must be maximized or minimized without violating any physical restraints or mission criteria. The remaining equations are solved in the usual manner.

CHAPTER III

DESCRIPTION OF THE DESIGN PROCEDURE

This chapter contains an outline of the design procedure, a list of its advantages, and three simple examples illustrating the use of the procedure.

Outline of the Design Procedure

The seven steps of the design procedure are outlined as follows:

1. Generalize a typical design problem. Select a typical design problem and generalize it into a set of design problems by permitting most or all of the specifications to vary. The resulting variables should consist of the desired performance, all important aspects of the operating situation, and perhaps one or more design characteristics. Nondimensionalize the variables to obtain a preliminary set of mission parameters. Specify the general design objective, the nondimensional optimization criterion, and all (fixed and dimensionless) mission criteria which are to be imposed upon the set of design missions.

2. Determine possible design forms. Sketch a wide variety of design forms, each of which may satisfy one of the many possible design missions. Conduct a brief analysis to find the most typically representative forms.

3. <u>Introduce physical relationships</u>. Determine all distinct physical phenomena which relate to the design problem. Develop dimensional relationships which associate the design problem specifications with the dimensional design variables and the optimization criterion; the physical phenomena can be used as a guide. Determine whether two or more values for any design variable result for a given design problem; if so, regions will overlap in mission space. Nondimensionalize each relationship to obtain dimensionless groupings of the design problem specifications. Reduce the nondimensional relationships, if possible, to the design equations and the optimization criterion.

4. <u>Select the mission and design parameters</u>. Select the sets of independent (nondimensional) mission and design parameters which appear most useful from the dimensionless groupings, the preliminary set of mission parameters, and sketches of design forms. Specify a preliminary coordinate system for mission space consisting of mission parameters and a preliminary coordinate system for design space consisting of design parameters. The two spaces are considered to be multidimensional Cartesian spaces. A single point in mission space determines a specific design mission and can be expressed as a series of numbers representing the values of the ordered coordinates. Similarly, a single point in design space determines a specific design form.

5. <u>Specify the mapping criteria and the design equations</u>. As a result of selecting the mission and design parameters, the mission criteria and the optimization criterion Q should be

rewritten, if necessary, so they are functions of the new parameters.
Also, the expressions should be simplified, if possible.

At this stage of the design process, it is possible to
specify the design equations. Sometimes Q can be utilized to
simplify the design equations by permitting an inequality sign to
be removed.

6. <u>Select a sequence of subspaces to map from mission space</u>.
The subspaces are to be mapped into design space, where mapping is
defined as the process of associating a design form with a point
in a region of mission space. The associated design form must
optimally satisfy the mapping criteria which consist of the optimization criterion and the mission criteria. In general, the best
mapping sequence consists of proceeding from subspaces which
represent the most simple and familiar design missions to those
which represent the more complex and least familiar design missions.

A possible mapping sequence consists of mapping one or two
simple points from mission space, each of three selected coordinate
axes, the three coordinate planes formed by these coordinate axes,
the three-dimensional subspace formed by the three coordinate planes,
and finally, other three-dimensional subspaces. Selecting each new
subspace to border on previously-mapped subspaces may aid signifi-.
cantly in simplifying the mapping process. Perhaps the simplest
point to map is the point whose coordinates are either zero or
infinity, depending upon which value effectively eliminates that
particular parameter from being significant. The associated design
form is generally the ideal design form since its design is less

limited by the mission parameters than any other form corresponding
to the selected subspace of mission space.

7. <u>Map from mission space to design space</u>. Consider each
subspace of mission space separately and in the order of the selected
sequence. Before conducting each mapping, determine whether or not
the selected subspace corresponds to more than one family of design
forms. If it does, the subspace will consist of more than one region,
each of which maps into a distinct family of design forms by means
of a distinct set of mapping relations. The mapping relations for
each region consist of the appropriate design equations plus a
certain number of relations which are derived from the optimization
criterion. This certain number is equal to the number of design form
parameters which are to be determined minus the number of design
equations.

During the process of establishing the mapping relations,
some of the mission parameters may be found to combine into new
mission parameters; if the use of a new set of mission parameters
appears to simplify the mapping relations, then adopt the new set
of parameters as the new mission space coordinates. Similarly, the
designer may find that the design form description is simplified by
changing the design parameters.

The mapping result can be illustrated by a series of graphs
which represent various one-, two-, and three-dimensional subspaces of
mission space on which are superimposed the region boundaries, if
any, the value of the associated optimization criterion, and the
value of some or all of the associated design form parameters.

The final set of mission space and design space coordinates will, in general, be the dimensionless parameters and graph coordinates which best serve to classify the various design missions and design forms, respectively. The design families corresponding to the various regions, if any, can be classified by the phenomena responsible for establishing the region.

Advantages of the Design Procedure

Some of the advantages resulting from the use of the design procedure are:

Design form variation. The mapping of points from various regions of mission space into design space provides the designer with an excellent understanding of the diversity in possible design forms and the reasons for such diversity.

Simplification. The design procedure helps simplify the treatment of design missions which include many variables or many optimization parameters. Also, the procedure aids in organizing a complex design mission and in determining where to begin the design process.

Organization of information and research studies. The use of this procedure aids in organizing information pertaining to the design field. Regions in mission space may be found where information is lacking and where a research study or invention is needed. The various mission parameters and design parameters resulting from use of the procedure can be utilized as experimental variables in research studies. Also, the design procedure can be modified and used to solve a research problem.

Clarification. The mapping process clearly shows that the best design in one design mission may not be the best in another. Consequently, questions of whether one design form is better than another can often be clarified by use of this procedure to show that each may be best for different design missions, or for similar missions with a different optimization criterion or different fixed mission criteria.

Design time and design quality. Once a generalized study of this type has been conducted, the time required to solve a specific design problem is significantly reduced. Furthermore, the resulting design form solution may be more acceptable than the usual solution since it may have been based on a more rational approach in which more variables were considered.

Scaling. The use of a nondimensional approach permits numerous design forms and design missions to be collapsed into relatively few parameters. These parameters permit broad scaling of the design forms and associated design missions.

Classification. The mission parameters, design parameters, and mission space regions which result from the use of the design procedure serve to classify design forms and their families.

The Design of Circular Tubes Subjected to External Pressure

This design example is the first of three simple examples presented in this chapter to illustrate the design procedure. Although no new technical information results from these examples, they serve to illustrate the methods of generalizing a design mission,

solving a set of design missions, and graphically presenting the

solutions. Any questions not answered by these examples may be

found answered in Chapter IV. The following steps for solving the

tube design problem are the same as those listed in the outline of

the design procedure:

1. <u>Generalize a typical design problem</u>. The typical design

problem selected for this example consists of determining the thick-

ness of an infinitely long circular tube which is two feet in

diameter and is submerged in sea water at a depth of 100 feet. The

tube is constructed from 7075 heat treated aluminum, and is filled

with air under atmospheric pressure. The weight of the tube is to

be minimized. Any permanent yielding or buckling of the tube is to

be prevented. The factor of safety is 1.5, assuming that the depth

pressure is steady.

This design problem is generalized by permitting the follow-

ing items to vary: (a) tube radius R, (b) pressure difference

across the tube wall times the safety factor p, (c) proportional

stress limit of the tube material in compression f_p, (d) weight

density of the tube material γ_t, (e) elliptical out-of-roundness

of the tube e measured as the maximum deflection from the desired

circle, and (f) modulus of elasticity of the tube material E.

Notice that the variable p includes the effect of variable depth,

fluid density, internal pressure, and safety factor.

The above variables are the set of mission variables which

consist of R, p, f_p, γ_t, e, and E. By inspection, four independent nondimensional mission parameters can be formed.[1] One possible set

of mission parameters is p/f_p, f_p/E, e/R, and $\gamma_t R/p$.

The (fixed) mission criteria are: (a) the tube cross section

is constant, (b) the tube is uniformly thick, (c) the tube thickness-

to-radius ratio t/R is much less than one, (d) there is no pressure

variation around the tube, and (e) the length-to-radius ratio of

the tube is infinity. These mission criteria were selected to

simplify the problem. All of the mission criteria could have been

considered as variables, in which case they would have been mission

parameters, and the set of mission criteria would have been empty.

The nondimensional optimization criterion Q is the non-

dimensional tube weight, one form of which is the (approximate) tube

weight per unit length $2\pi\gamma_t Rt$ (where t is the tube thickness)

divided by pR, which results in

$$Q = 2\pi\gamma_t t/p \qquad\qquad (10)$$

where Q is to be minimized. The general design objective is to

determine the cross-sectional form of the pipe so that no permanent

yield occurs.

2. <u>Determine possible design forms</u>. Possible design forms

are the following:

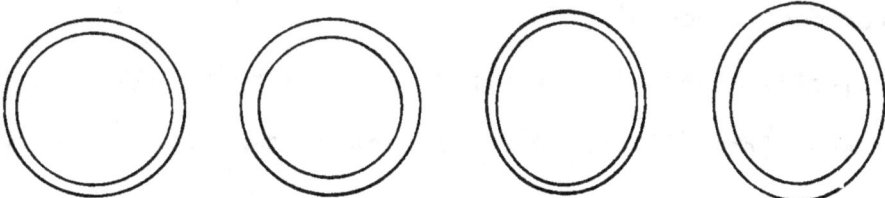

[1] See the section "Nondimensional Parameters" in Chapter IV
for information on the pi theorem which can be used to
uniquely determine the quantity of nondimensional parameters.

The only design variable is clearly seen to be the tube thickness t since the radius and out-of-roundness are mission variables and therefore specified in a given design mission.

    3.   <u>Introduce physical relationships</u>. The physical phenomena pertaining to this problem are compressive stress and buckling. There are no effects of gravity other than to provide tube weight which has already been included as the optimization criterion Q; consequently, no additional physical relationships regarding gravity and γ_t exist.

    The relationship for the compressive stress of a circular tube can be found from the tube geometry, and is

$$(\text{compression}) \quad f_p \geq \frac{pR}{t} \tag{11}$$

where pR/t is the compressive stress which must be less than or equal to f_p to prevent permanent yielding.

    The relationship for the prevention of buckling of an infinitely long tube is obtained from Timoshenko (11), as

$$(\text{buckling}) \quad p \leq \frac{E}{4(1-\mu^2)} \left(\frac{t}{R}\right)^3 = \frac{E}{3.64} \left(\frac{t}{R}\right)^3 \tag{12}$$

where μ is Poisson's ratio and has been set equal to 0.3 which is valid for most types of structural material.

    Finally, the relationship for the out-of-round (elliptical) tube is seen from (11) to be neither pure compression nor pure buckling, but a combination of compression and bending stress where

$$\text{(out-of-round)} \quad f_p \geq \frac{pR}{t} + \frac{6pRe/t^2}{1 - \frac{4p(1-\mu^2)R^3}{Et^3}} = \frac{pR}{t} + \frac{6pRe \cdot t^2}{1 - \frac{3.64pR^3}{Et^3}} \quad (13)$$

Nondimensionalizing Equations 11 to 13,

$$\text{(compression)} \quad \frac{t}{R} \geq \frac{p}{f_p} \quad (14)$$

$$\text{(buckling)} \quad \frac{t}{R} \geq 1.54 \left(\frac{p}{E}\right)^{1/3} \quad (15)$$

$$\text{(out-of-round)} \quad \frac{f_p}{p} \geq \frac{R}{t} + \frac{6 \left(\frac{R}{t}\right)^2 \left(\frac{e}{R}\right)}{1 - 3.64 \left(\frac{p}{E}\right) \left(\frac{R}{t}\right)^3} \quad (16)$$

The design equations for the different mappings are Equations 14 to 16, and Equation 10 is the optimization criterion. Since three different relationships exist for t/R, mission space will split into three different regions.

4. <u>Select mission and design parameters</u>. After inspecting Equations 14 to 16 and the design forms sketched in Step (2), the best design parameter appears to be t/R[1]. A reasonable set of mission parameters, after inspecting the preliminary set of Step (1) and Equations 14 to 16 appear to be p/f_p, p/E, and e/R.

---

[1] An alternate parameter is t/e, but this would require t/R in Equations 14 and 15 to be treated as t/e times e/R, which is an unnecessary complication.

Since no physical relationship, other than the one which became the optimization criterion, was found for the parameter $\gamma_t R/p$, it should not be included as a mission parameter; notice that a change in its value has no effect on the design form parameter t/R.

Summarizing, mission space may be looked upon as being three-dimensional and consisting of the coordinates

$$m_1 = p/f_p \qquad m_2 = p/E \qquad m_3 = e/R$$

Design space consists only of the coordinate

$$d_1 = t/R$$

5. <u>Specify the mapping criteria and the design equations.</u>
In view of the new design parameter t/R, Equation 10 can be rewritten as

$$Q = 2\pi \left(\frac{\gamma_t R}{p}\right) \left(\frac{t}{R}\right) \tag{17}$$

Since Q is to be minimized and $\gamma_t R/p$ is specified for a given design mission, Equation 17 shows that t/R should be minimized.

In order to minimize t/R, the inequality signs should be removed from the design equations, Equations 14 to 16, which become

$$\text{(compression)} \quad \frac{t}{R} = \frac{p}{f_p} \tag{18}$$

$$\text{(buckling)} \quad \frac{t}{R} = 1.54 \left(\frac{p}{E}\right)^{1/3} \tag{19}$$

$$\text{(out-of-round)} \quad \frac{f_p}{p} = \frac{R}{t} + \frac{6 \left(\frac{R}{t}\right)^2 \left(\frac{e}{R}\right)}{1 - 3.64 \left(\frac{p}{E}\right) \left(\frac{R}{t}\right)^3} \tag{20}$$

6. <u>Select a sequence of subspaces from mission space.</u> A possible sequence of subspaces to be mapped is: (a) $m_1 = m_2 = m_3 = 0$, (b) $m_2 = m_3 = 0$, (c) $m_1 = m_3 = 0$, (d) $m_3 = 0$, and (e) $m_3 = 0.005$. Note that $m_3 = 0$ in all subspaces except the last, so the mission parameter e/R, which specifies the out-of-roundness, is considered only in Subspace (e). The value of e/R = 0.005 selected for m_3 in Subspace (e) is typical of a possible out-of-roundness resulting from certain types of manufacturing methods.

7. <u>Map from mission space to design space.</u> The first subspace to be considered is Subspace (a) which consists of $m_1 = m_2 = m_3 = 0$. This subspace is a point. Equations 18 and 19 show that t/R = 0.[1] In other words, the optimized tube corresponding to the design mission described by the point $m_1 = m_2 = m_3 = 0$ has zero thickness. This design form is unquestionably ideal from the viewpoint of minimizing tube weight; however the design mission is not a practical one.

Subspace (b) consists of variable $m_1 = p/f_p$ where $m_2 = m_3 = 0$. Equation 18 shows that the mapping relationship is simply $t/R = p/f_p$. In other words, the ratio t/R is directly proportional to p and inversely proportional to f_p. There are no critical values or boundaries in this subspace of mission space.

---

[1] Equation 20 is not utilized since e/R = 0.

Similarly, Equation 19 shows that the mapping relationship for Subspace (c) is $t/R = 1.54 \ (p/E)^{1/3}$. In this case, buckling is critical and only the values of p and E are important. The compressive stress limit f_p is not significant. No boundaries exist in this subspace.

Subspace (d) represents the first set of design missions of practical value, since only the parameter $m_3 = e/R$ is zero and m_1 and m_2 are variable. In this mapping, both Equations 18 and 19 are required. A boundary will exist in this section of mission space since Equations 18 and 19 provide two different values for t/R, indicating that two regions exist. One region corresponds to designs where the compressive stress is critical, and the other corresponds to designs where buckling is critical. At any one point in Subspace (d), the largest of the two values of t/R calculated from Equations 18 and 19 represents the most critical condition since failure would occur if t/R were equated to the smaller value. The line in Subspace (d), which corresponds to the case when Equations 18 and 19 provide the same value of t/R, is the desired boundary since it represents the cross-over from a compression-limited design form to a buckling-limited design form. Equating t/R of Equations 18 and 19 results in

$$[\text{boundary, Subspace (d)}] \quad \frac{p}{f_p} = 1.54 \ (\frac{p}{E})^{1/3} \qquad (21)$$

In view of the above discussion, Equation 18 pertains to the region $p/f_p \geq 1.54 \ (p/E)^{1/3}$ and Equation 19 pertains to the region $p/f_p \leq 1.54 \ (p/E)^{1/3}$. The result of this mapping is illustrated in

Figure 4 which is a graph of the (p/f_p) vs. (p/E) space upon which the boundary and the associated values of t/R are super'mposed.

To find the tube form having least weight for a specific mission, one must first calculate both p/f_p and p/E and then find the value of t/R which corresponds to that point in Figure 4. The solution to the specific design mission presented earlier can now be readily found. The values of f_p and E for 7075 heat-treated aluminum are 73,000 psi and 10.4×10^6 psi, respectively. The pressure differential across the tube multiplied by the safety factor is p = (64) (100) (1.5)/144 = 66.7 psi. Consequently, $p/E = 6.5 \times 10^{-6}$ and $p/f_p = 9.3 \times 10^{-4}$. From Figure 4 it is seen that buckling is critical and t/R $\doteq$ 0.03. A more accurate value for t/R of 0.0287 is obtained by using Equation 19. The thickness of the tube which has a one-foot radius is therefore 0.0287 feet or 0.344 inches.

In case the designer would rather present the mapping result in a graph which provides a more accurate design solution than Figure 4, he could change the form of mission parameter m_2 from m_2 = p/E to $m_2 = 1.54 \, (p/E)^{1/3}$. The result is Figure 5 which needs no scale since m_1 and m_2 are each t/R. The use of Figure 5 is equivalent to selecting the largest value of t/R resulting from Equations 18 and 19.

The designer may desire to find a different graphical presentation which is easier to use than Figure 5 and yet provides a more accurate result than Figure 4. One way of doing this is to construct a series of graphs, each of which pertains to a single type of

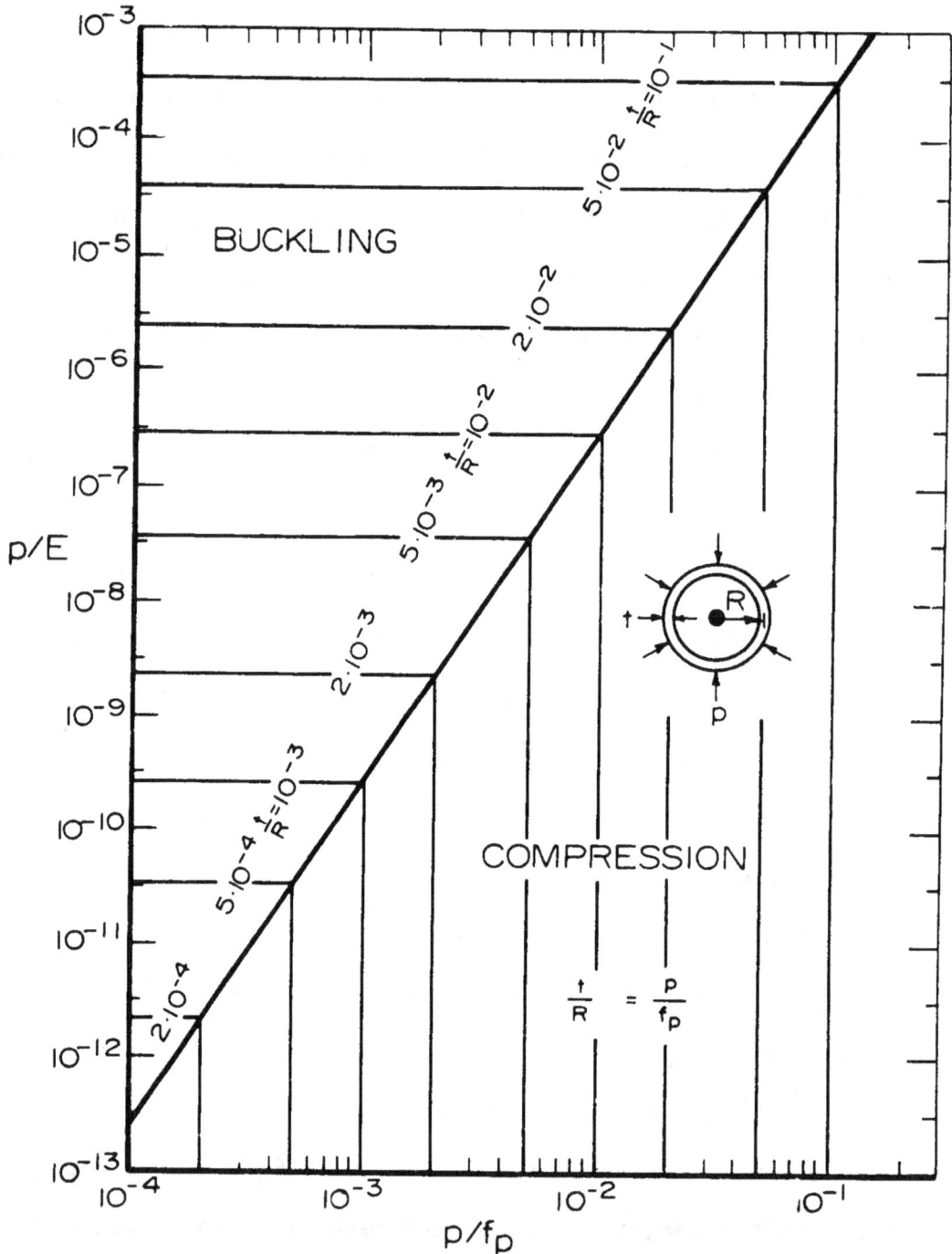

Figure 4 – Design of circular tubes subjected to external pressure, $(p/f_p,\ p/E)$ space

54

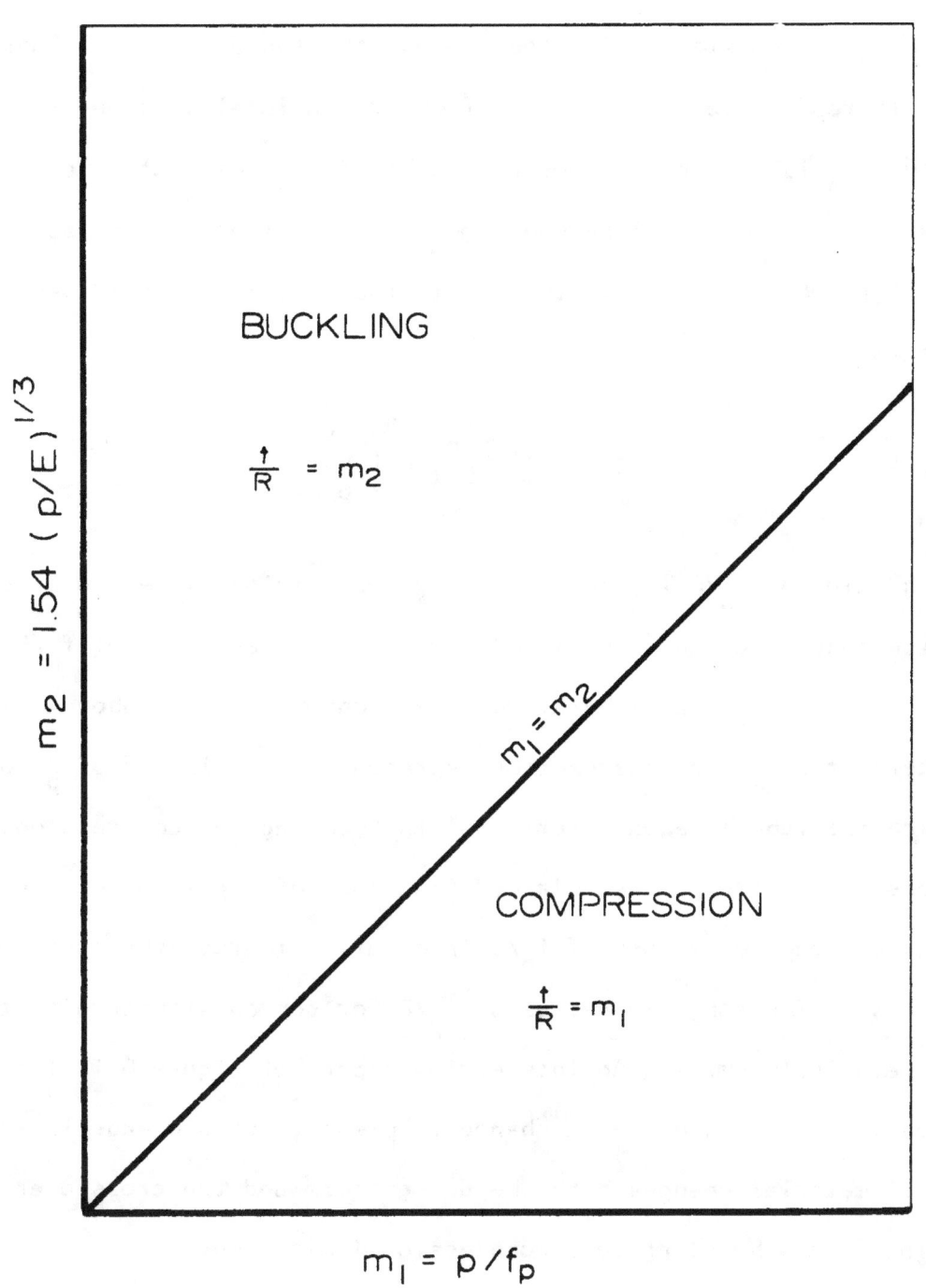

Figure 5 – Design of circular tubes, $[p/f_p, 1.54 \ (p/E)^{1/3}]$ space

structural material, in which t/R is plotted as a function of p/f_p with f_p/E = constant. In other words, the two-dimensional Subspace (d) is represented by a series of one-dimensional sections in which only p/f_p is variable. However, rather than construct a series of graphs, they could all be superimposed on a single graph, as shown in Figure 6. In order to construct Figure 6, Equation 19 was transformed into

$$\frac{t}{R} = 1.54 \left(\frac{f_p}{E}\right)^{1/3} \left(\frac{p}{f_p}\right)^{1/3} \tag{22}$$

Each value of f_p/E is represented by two straight lines in Figure 6 which consist of the line labeled with the given value of f_p/E and that portion of line labeled "compression" which lies above their intersection. The intersection represents the value of p/f_p for which the tube is equally critical in buckling and compression. The scale of t/R can be enlarged from that of Figure 6 to cover a smaller range of values of f_p/E in order to improve the accuracy of its use. The range in values of f_p/E for common structural materials is relatively small. An interesting aspect of Figure 6 is that it clearly shows how either a change in pressure or a change in structural material changes both the design form and the cross-over point from a buckling to a compression limitation.

Still another form of graphical presentation is shown by Figure 7 in which p/f_p is graphed against f_p/E with the corresponding values of t/R superimposed. A horizontal line represents a given structural material. This graph has all of the advantages of Figure 6

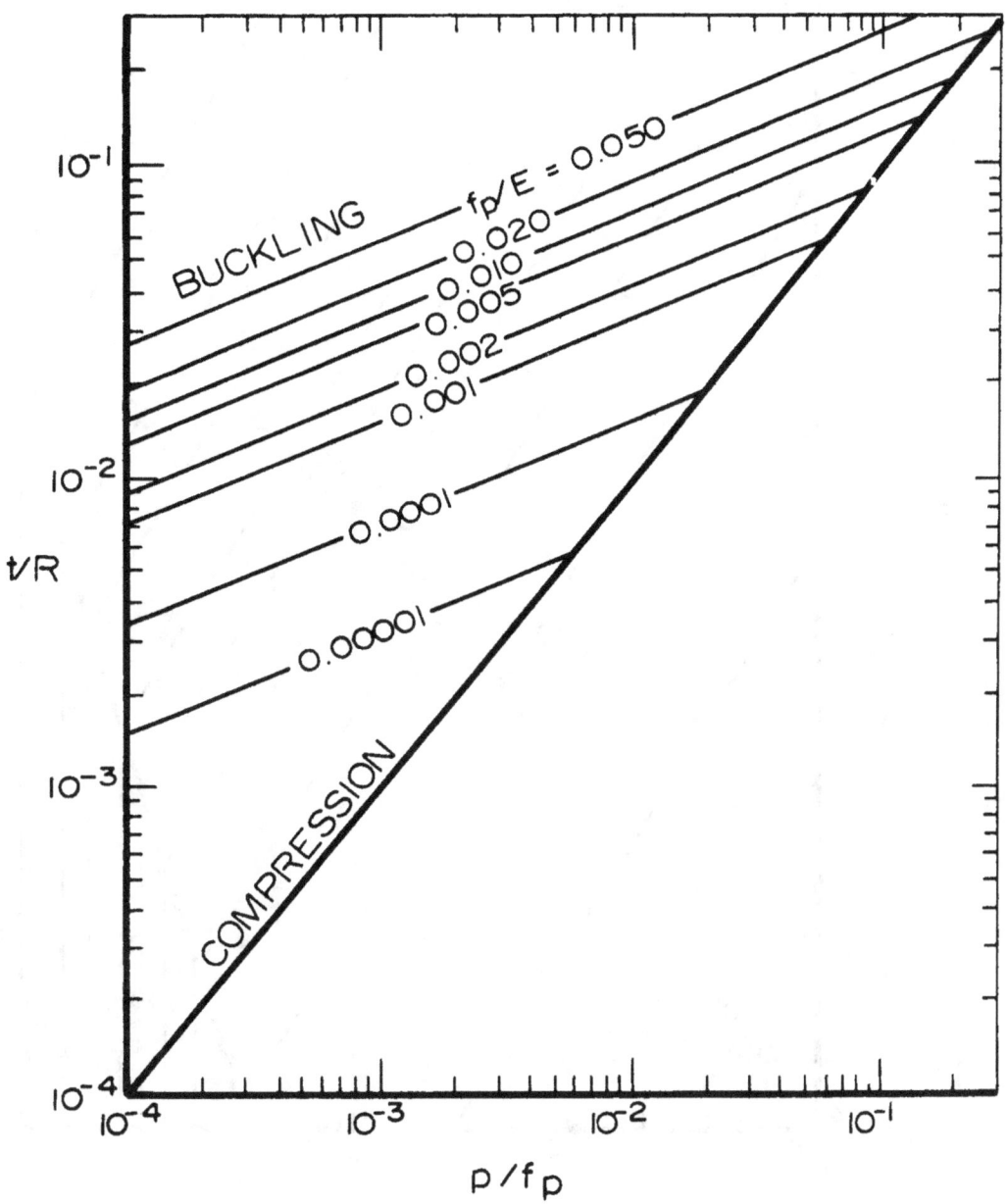

Figure 6 - Design of circular tubes, p/f_p space

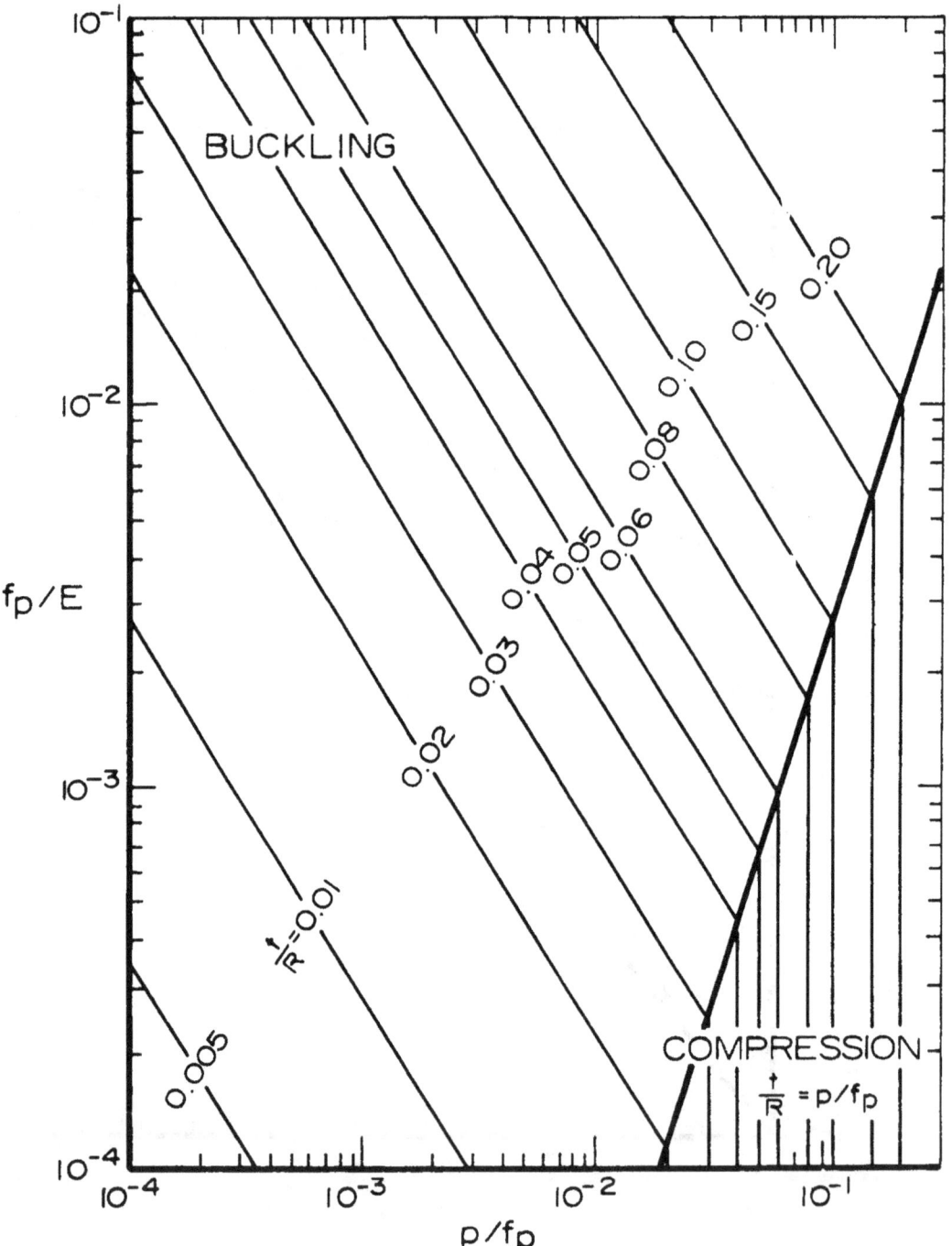

Figure 7 - Design of circular tubes, $(p/f_p, f_p/E)$ space

in addition to the fact that it can be used as a starting point for graphing the entire three-dimensional section of mission space which 's discussed next.

The three-dimensional section, Subspace (e), is mapped into design space by means of Equation 20. Rewriting Equation 20,

$$\text{(out-of-roundness)} \quad \frac{f_p}{E} = \left(\frac{t}{R}\right)^2 \frac{[\frac{t}{R} + 6 \frac{e}{R} - \frac{f_p}{p} \left(\frac{t}{R}\right)^2]}{3.64 \left(\frac{p}{f_p} - \frac{t}{R}\right)} \quad (23)$$

Equation 23 is plotted in Figure 8 where e/R has been equated to 0.005. The boundary for the case when e/R = 0 is shown by the dotted line. Notice that no boundary exists in Figure 8 since only one mapping relation is relevant. If Figure 8 is placed directly above and parallel to Figure 7, the two graphs would represent a portion of the three-dimensional space of Subspace (e) where e/R is the coordinate pointing upward and normal to the planes of the graphs. The lines of t/R in Figure 8 merge into the lines of t/R in Figure 7 as e/R approaches zero. The boundary shown in Figure 7 appears only when e/R is exactly zero. The boundaries do not always disappear so abruptly in a new dimension of mission space, as shown by some of the design problems presented later.

The design form family of infinitely long circular tubes subjected to external pressure can be classified by thickness-to-radius ratio. The associated design missions can be classified by p/f_p, if the structual material is fixed; otherwise, f_p/E must be added. The out-of-roundness criterion e/R may serve either as an

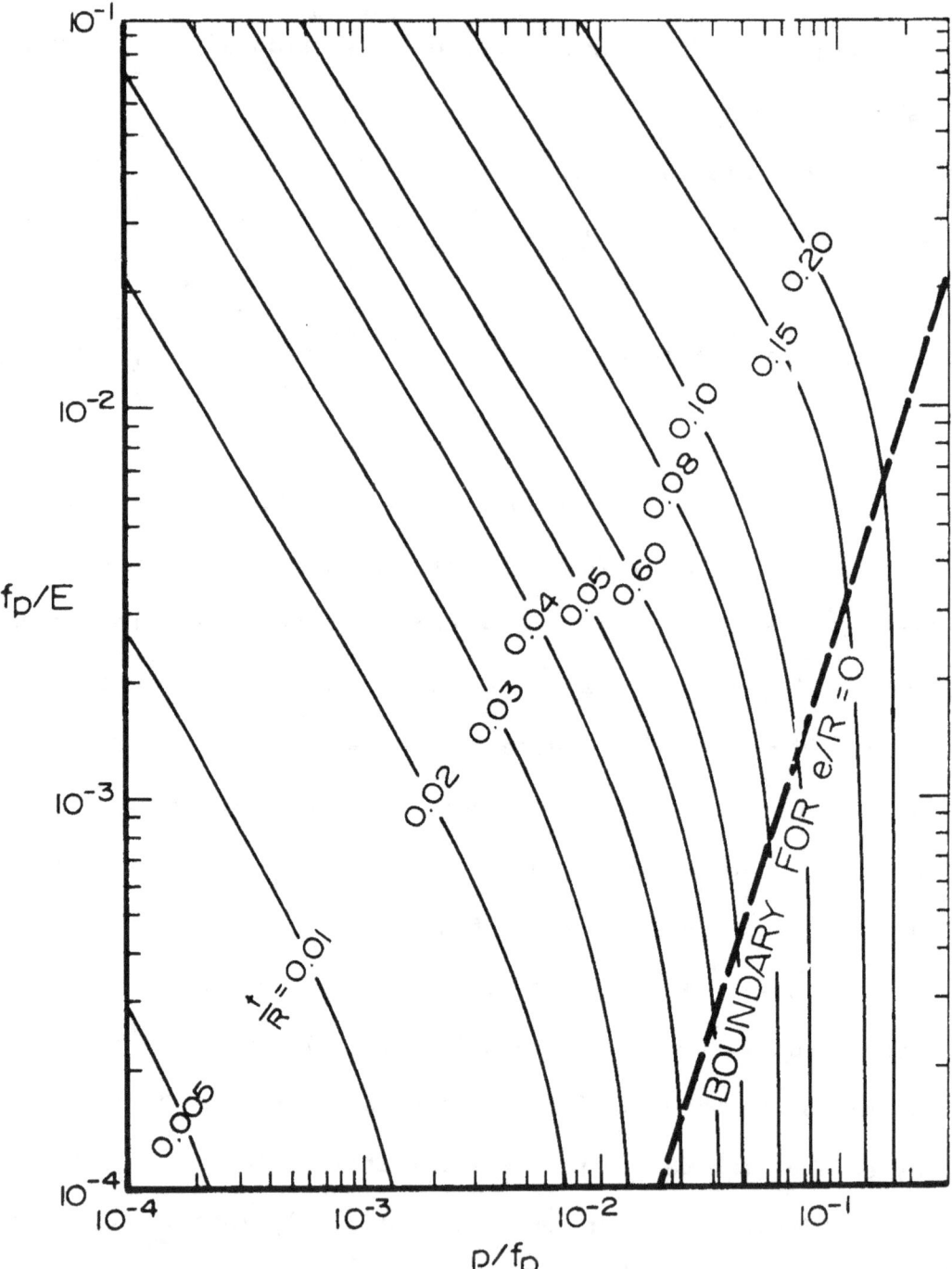

Figure 8 - Design of circular tubes, $(p/f_p, f_p/E, e/R)$ space, $e/R = 0.005$

additional parameter for classifying the design form or for classifying the design mission, depending upon the nature of the design
mission under consideration.

Although this first design example is a relatively common
one and the graphs which illustrate the design result appear quite
ordinary, it is unlikely that the reader has seen some of these types
of graphs before. This result demonstrates another feature of the
design procedure, namely that new and useful means of illustrating
design solutions may result.

The Design of Cylindrical Columns Loaded in Compression

This second design example is closely related to the first,
since it is also a structural problem in which the design may fail
under either pure compression or buckling. This example is shortened
by leaving out those steps of the design procedure which can often
be conducted mentally in order to illustrate that the procedure is
basically simple.

A typical column problem is generalized by permitting the
following design specifications to vary: (a) design load W which
includes the factor of safety, (b) column length ℓ, (c) proportional
stress limit in compression f_p, (d) modulus of elasticity E,
(e) cross-sectional shape, and (f) end conditions of the column.
The fixed mission criteria are: (a) the design load is steady,
(b) the load is applied exactly at the structural center of the
column, (c) the column weight is small relative to the applied load.
The cross-sectional area of the column is to be minimized.

The design objective is to determine the column form.

The design problem variables are W, ℓ, f_p, E, cro.s-sectional shape, and the end conditions. A possible set of nondimensional mission parameters are $W/f_p\ell^2$, f_p/E, and two (as yet unestablished) parameters which represent the cross-sectional shape and the end conditions.

The physical relationships (11) are

(compression) $\qquad W \leq f_p A \qquad\qquad\qquad\qquad$ (24)

(buckling) $\qquad W \leq \dfrac{\pi^2 EI}{n^2 \ell^2} = \dfrac{\pi^2 E r^2 A}{n^2 \ell^2} \qquad\qquad$ (25)

where A = cross-sectional area, $I = r^2 A$ = minimum area moment of inertia, r = radius of gyration, and n represents the end conditions where n = 0.5 for cantilevered ends, n = 0.7 for one cantilevered and one hinged end, n = 1.0 for two hinged ends, and n = 2 for one cantilevered end and one free end.

In view of the optimization criterion (that the cross-sectional area is to be minimized) the inequality signs are removed from Equations 24 and 25. Nondimensionalizing Equations 24 and 25 (keeping in mind the two preliminary mission parameters, the fact that A is not specified, and the need for both a cross-sectional form parameter and a design form parameter) leads to the following:

(compression) $\qquad \dfrac{W}{f_p \ell^2} = \dfrac{A}{\ell^2} = \left(\dfrac{A}{r^2}\right)\left(\dfrac{r^2}{\ell^2}\right) \qquad\qquad$ (26)

(buckling) $\qquad \dfrac{W}{f_p \ell^2} = \dfrac{\pi^2 E r^2 A}{n^2 f_p \ell^4} = \dfrac{\pi^2}{n^2}\left(\dfrac{E}{f_p}\right)\left(\dfrac{r^4}{\ell^4}\right)\left(\dfrac{A}{r^2}\right) \qquad$ (27)

where A/r^2 is the desired mission parameter which describes the cross-sectional shape, and r^2/ℓ^2 is the desired design form parameter which describes the column slenderness and which permits the column form to be determined. Equations 26 and 27 show that only two, rather than four, mission parameters are required. These are

$$m_1 = \frac{W}{f_p \ell^2} \cdot (\frac{r^2}{A}) \tag{28}$$

$$m_2 = \frac{f_p n^2}{E} \tag{29}$$

Substituting Equations 28 and 29 into Equations 26 and 27, the two design equations become

(compression) $\qquad m_1 = (\frac{r}{\ell})^2 \tag{30}$

(buckling) $\qquad m_1 m_2 = \pi^2 (\frac{r}{\ell})^4 \tag{31}$

Equations 30 and 31 show that the m_1 vs. m_2 space must consist of two regions separated by a boundary since two values of r/ℓ result, the largest of which determines the equation which dominates in a given region. The boundary line between the regions is obtained by setting r/ℓ equal in Equations 30 and 31, which gives

(boundary) $\qquad m_2 = \pi^2 m_1 \tag{32}$

Figure 9 is a graph of the m_1 vs. m_2 space with the boundary and the associated values of r^2/ℓ^2 superimposed. Equations 30 to 32 were utilized to construct the graph. Since ℓ is known for a given

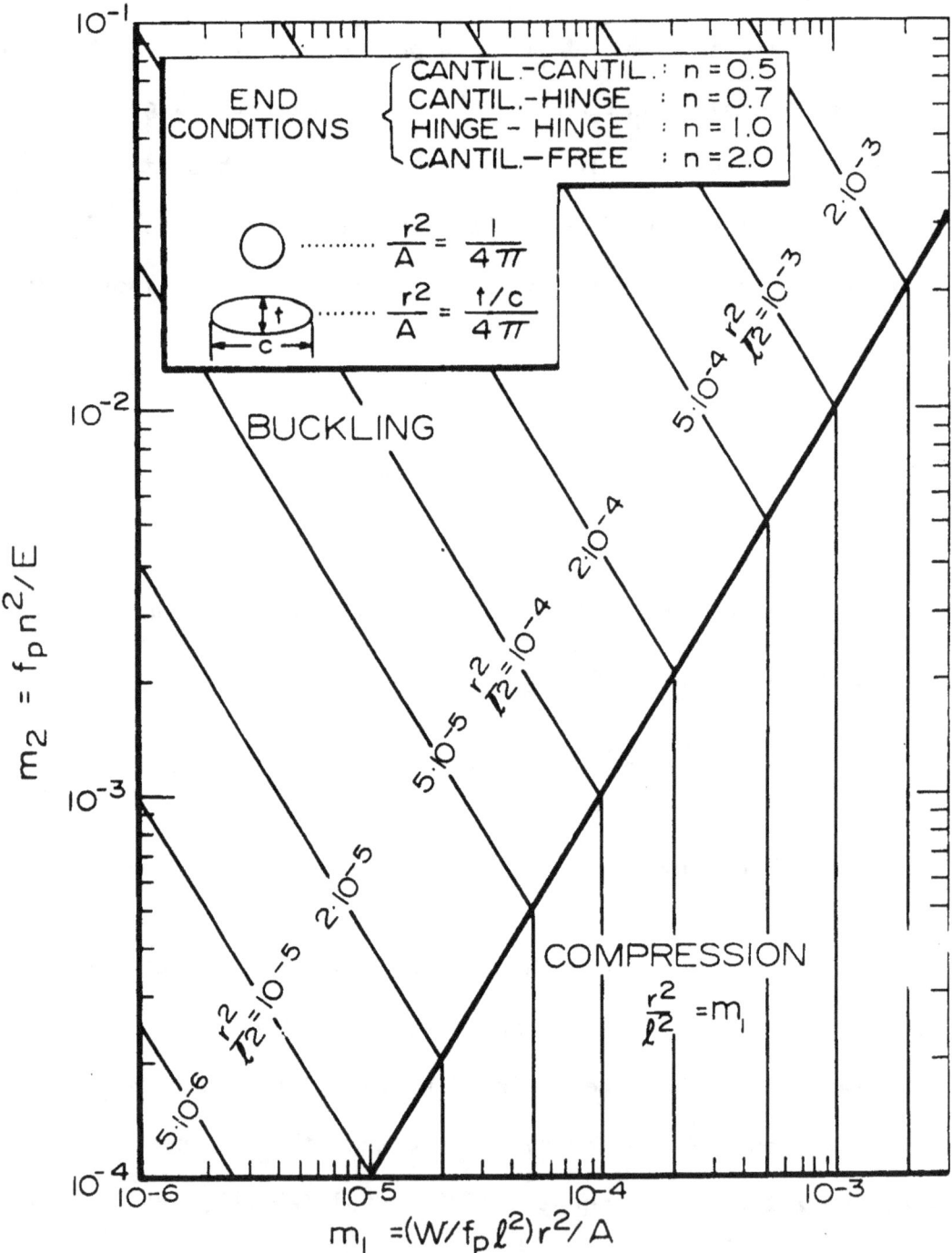

Figure 9 — Design of cylindrical columns loaded in compression

mission, r can be calculated from the resulting value of r^2/ℓ^2.
Similarly, r^2/A is known, so A can be evaluated. Consequently, the
column size is known. For convenience, values of r^2/A are listed in
Figure 9 for a circle and an ellipse.

The results of this example show that column form can be
classified by r/ℓ and r^2/A. The design mission can be classified by
$(W/f_p\ell^2)(r^2/A)$ and $f_p n^2/E$, and the column support by n.

Economic Example

In order to illustrate the application of the design procedure
to a field outside of engineering, an economic example is selected.
For this example, the following words in the design procedure must
be changed to the word in parenthesis: design mission (mission),
physical relationships (mission relationships), design form (solution),
design space (solution space), design parameters (solution parameters),
and design equations (solution equations).

The selected typical economic problem concerns an item that
is sold retail for $10. The item can be manufactured by one of three
methods. The first method entails no tooling, and the cost of
material is $1 per item and the fabrication cost is $4 per item.
The second method is semi-automated, but due to some wastage of
material, the cost of material is $1.20 per item and the fabrication
cost is $2 per item. The third method is highly automated, and the
cost of material and fabrication cost are each $1 per item. The
tooling for the second method costs $5,000, and the tooling for the
third method costs $20,000. Approximately 10,000 items are to be

produced. The gross profit (i.e., the difference between the total selling price and the total manufacturing cost) is to be maximized. The problem is to determine which manufacturing method is best.

Generalized mission. This economic problem can be generalized in several ways, depending upon which factors are made variable. One type of generalized mission is to consider the following specifications to be variable: tooling cost A_1; cost of material per item A_2; fabrication cost per item A_3; retail price per item A_4; and number of items sold N. The general mission objective is to determine the gross profit. One possible set of nondimensional mission parameters is α, β, γ, and N, where $\alpha = A_1/A_4$, $\beta = A_2/A_4$, and $\gamma = A_3/A_4$.

The gross profit is easily seen to be

$$\text{gross profit} = A_4 N - A_1 - N(A_2 + A_3) \tag{33}$$

where the total manufacturing cost of the items sold is $A_1 + N(A_2 + A_3)$. Let the nondimensional gross profit p be the gross profit divided by the total cost of items sold. From Equation 33,

$$p = \frac{A_4 N}{A_1 + N(A_2 + A_3)} - 1 = \frac{1}{\frac{\alpha}{N} + \beta + \gamma} - 1 \tag{34}$$

Because of the way in which the generalized mission was set up, there is nothing to optimize; consequently, there is no optimization criterion. The mission solution is p and, in view of Equation 34, can be expressed directly as a function of the mission parameters m_1 and m_2 as

$$P = \frac{1}{m_1 + m_2} - 1 \tag{35}$$

where $m_1 = \alpha/N = A_1/A_4 N$, $m_2 = \beta + \gamma = (A_2 + A_3)/A_4$, and p is the gross profit divided by the manufacturing cost of goods sold.

The solution is graphed in Figure 10 which shows p as a function of m_1 and m_2. Either Figure 10 or Equation 35 can be used to solve any of a variety of related economic problems.

The respective values of m_1 and m_2 in the original economic problem for the first, second, and third manufacturing methods are $m_1 = 0$, $m_2 = 0.500$; $m_1 = 0.050$, $m_2 = 0.320$; and $m_1 = 0.200$, $m_2 = 0.200$. The respective values of p are 1.00, 1.70, and 1.50. Consequently, the second method is best, since p is to be maximized.

The best classification parameters for this type of economic problem are m_1 and m_2, where m_1 = the tooling cost divided by the number of items sold and retail price per unit, and m_2 = the sum of the material cost and fabrication cost per item divided by the retail price per item.

Alternate generalized mission. Another way of generalizing the original economic problem is to consider the number of items sold N as the only mission parameter. The fixed mission criteria are the values of the parameters α and δ for each manufacturing method, where $\delta = \beta + \gamma$. The values of α and δ for each of three manufacturing methods are $\alpha_1 = 0$, $\delta_1 = 0.500$; $\alpha_2 = 500$, $\delta_2 = 0.320$; and $\alpha_3 = 2,000$, $\delta_3 = 0.200$. The subscripts refer to the manufacturing method. The parameter p in Equation 34 becomes the optimization

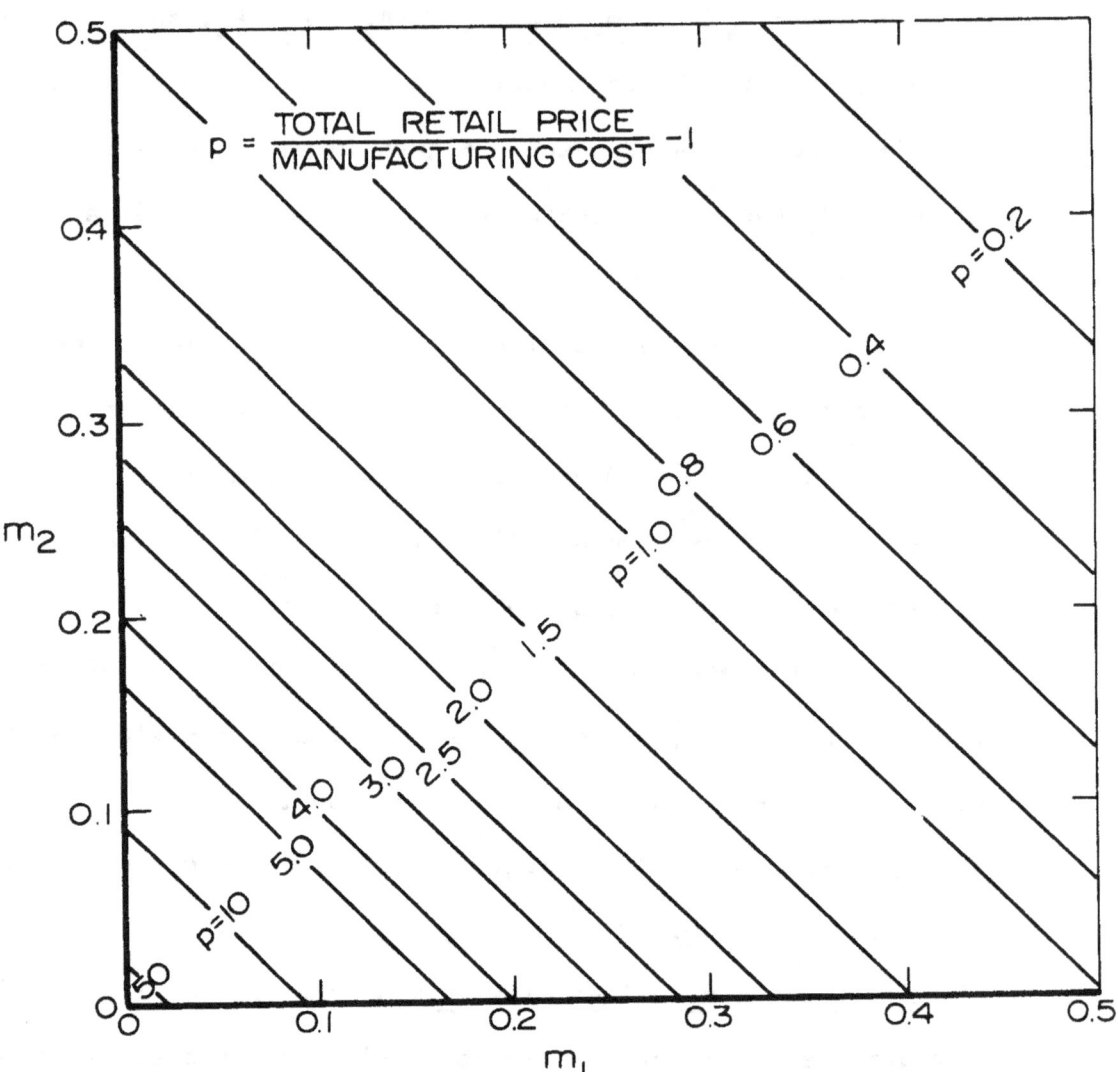

Figure 10 - Generalized economic problem No. 1

criterion Q where

$$p = Q = \frac{N}{\alpha + \delta N} - 1 \qquad\qquad (36)$$

and Q is to be maximized. The mission objective is to determine
the best manufacturing method as a function of N.

Notice that no meaningful economic relationships can be
written, other than the expression for Q. Consequently, the number
of solution equations is zero, so the information needed to solve
the problem must come from Q. The following are the values of Q as
a function of N for each of the three manufacturing methods, as
obtained from Equation 36 and the mission criteria:

$$P_1 = Q_1 = 1.00$$

$$P_2 = Q_2 = \frac{N}{500 + 0.320\ N} - 1 \qquad\qquad (37)$$

$$P_3 = Q_3 = \frac{N}{2,000 + 0.200\ N} - 1$$

Figure 11 is a graph of p versus N on which p_1, p_2, and p_3 of
Equations 37 are plotted. The solid line shows the optimum values
of p for any N, and also the best manufacturing method for any N.
The two intersection points of the three curves in Figure 11 are
the cross-over points from one manufacturing method to another and
are the region boundaries in the one-dimensional mission space
consisting of N. The regions in this mission space are the three
ranges of N corresponding to each of the manufacturing methods.

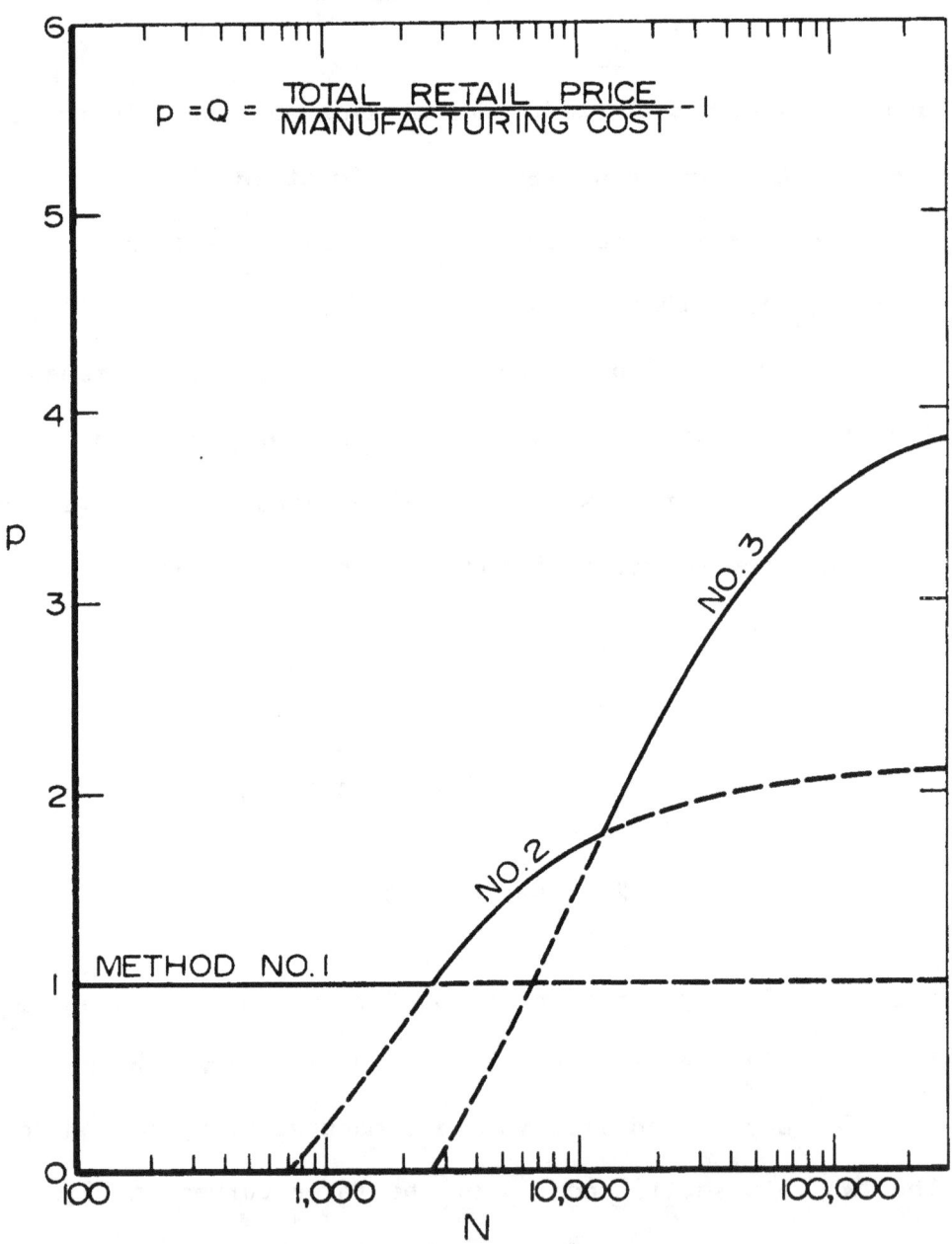

Figure 11 - Generalized economic problem No. 2

CHAPTER IV

DISCUSSION OF THE DESIGN PROCEDURE

The objectives of this chapter are to present additional
information on the design procedure, discuss the scaling of design
forms, discuss methods for applying the design procedure to fields
outside of engineering, describe how the procedure can be applied
to research studies, and comment on future developments in design
theory.

Methods for Nondimensionalizing Variables

The question of how to nondimensionalize a set of variables
is answered in the book by Kline called "Similitude Approximation
Theory" (12). Additional information is presented by Sedov (13).

Kline presents three basic methods for obtaining nondimensional parameters: (1) dimensional analysis, (2) the method of
similitude, and (3) the systematic use of the governing equations.
The approach suggested for this design procedure is a combination
of the first and third methods.[1]

---

[1] The method of similitude is not emphasized because, as
mentioned in (12), any result which it may provide is con-
tained in the results of the third method listed above.
The method of similitude consists fundamentally of deriving
nondimensional parameters from force ratios. It is often
a simple method to use, but it cannot provide some of the
combinations of parameters which result from the third
method and it cannot show the relative importance of the
parameters.

Dimensional analysis. This method of nondimensionalizing a set of quantities is relatively simple, although it requires knowledge of the physical situation to be most effective. It consists of the enumeration of the relevant physical quantities followed by the application of the pi theorem. The pi theorem was developed by Buckingham (14) and refined later by others. It consists of a relation which determines the number of all possible independent nondimensional groupings which can be formed from a set of dimensional quantities. According to (12), the number of independent groupings is simply the number of dimensional quantities minus k, where k is the smaller of the following: (a) the number of the relevant dimensions or (b) the maximum number of the given dimensional quantities that will not combine into any nondimensional form. Notice that k cannot exceed the number of relevant dimensions, and may be less than this number. For example, if the set of dimensional quantities consists of a speed, a length, and an acceleration, the pi theorem would predict $3 - 2 = 1$ nondimensional grouping since, of the three given quantities, only two dimensions are relevant, namely length and time. If instead, the set of dimensional quantities consists of a length, a volume, a pressure, and a force, the pi theorem would predict $4 - 2 = 2$ nondimensional groupings. The reason is, that although three dimensions are relevant, the maximum number of dimensional quantities that will not combine into a nondimensional form is two; therefore, two is the smaller of the items (a) and (b) given above, so $k = 2$.

Although the method of dimensional analysis is highly useful, it is unable to show the following: (a) whether all of the important variables were listed, (b) which of the many possible sets of resulting parameters is best, (c) the relative importance of each parameter, (d) which parameters can be neglected in a given situation, and (e) how the parameters might be combined into a smaller number of more useful parameters. Nevertheless, the prediction of the total number of independent parameters is highly important and useful.

Perhaps the best way of nondimensionalizing a set of dimensional quantities is by inspection, as recommended by Kline (12), because the designer can select the forms of the parameters which are most meaningful in view of his experience. As long as the total number of parameters agrees with the pi theorem, any set of parameters is equally as good as any other set. Whichever set is selected, however, it must contain each of the dimensional quantities at least once.

Governing relationships. The governing relationships of a design mission are all relationships which associate the dimensional design solution with the dimensional design problem variables, design problem criteria, and the optimization criterion. These relationships can be nondimensionalized to extract dimensionless groupings in the form of possible mission parameters, mission criteria, dimensionless optimization parameters, and design form parameters. The resulting set of dimensionless groupings are the necessary and sufficient parameters needed to describe the generalized mission and the resulting design forms.

The set of nondimensional parameters resulting from dimensional analysis is used as a guide in the design procedure for nondimensionalizing the governing relationships. The set of parameters resulting from dimensional analysis can also be used to help determine whether the set of governing relationships is complete and appropriate, even though the set may have been based on intuition.

The nondimensional form of the governing relationships shows which parameters can be combined, which parameters may be most useful, the relative importance of each parameter, and which parameter can be neglected in a given situation.

Selection of Mission Parameters

As shown by the design examples in Chapter III, the mission parameters are not necessarily unique; however, a certain set may be preferred. The preferred set depends upon the physical nature of the generalized design mission and upon the use of the generalized design mission solutions.

The use of the governing relationships. The mission parameters selected from the nondimensional groupings of the governing equations may differ depending upon how the governing equations are divided to nondimensionalize them, and whether the equations are combined. In many cases, combining or simplifying the nondimensional governing equations will provide a set of parameters which is the minimum possible number of independent mission parameters that can be used to define a given generalized design mission. Parameters of such a set are often combinations of parameters which might otherwise have been selected as individual mission parameters for mission space. An example of such a combination is shown in the design problem of Chapter V.

The use of design form solutions. Sometimes the design form solutions to a given generalized design mission suggest a new and simplified representation for mission space. An example is shown in the design problem of Appendix B where the solutions to the original generalized design mission suggest how three dimensions of mission space can be collapsed into a single dimension. This possibility was not apparent until the original generalized design mission was solved. The new mission parameter not only simplified the description of the generalized design mission and its solutions, but provided a new and more general parameter for classifying the design forms.

The use of physical knowledge. Knowledge of the general physical situation, theory, and experimental results of related research studies may show which nondimensional parameters are known to be most important. Such parameters can then be used as a guide in selecting mission parameters. Also, the use of the method of similitude described in (12) where force ratios are considered will provide mission parameters. Furthermore, as mentioned earlier in this chapter, physical knowledge of the relevant dimensional quantities followed by dimensional analysis will provide a set of mission parameters. Although the combined use of dimensional analysis and the governing equations is the preferred technique presented here

for selecting mission parameters, the other techniques may be
necessary in design problems where the governing equations are not
known.

Relationship of the mission parameters to the other parameters.
The mission parameters have certain properties which differentiate
them from the other parameters. These properties are discussed
below.

As mentioned in Chapter II, the mission parameters and the
mission criteria are the set of nondimensional criteria which comprise
the specifications for a given design mission. The only difference
between them is that the mission parameters are variables in a
generalized design mission and the mission criteria are fixed.
Therefore, all mission parameters and mission criteria are independent of each other and, together with a general design objective
and an optimization criterion, they completely specify any design
mission.

All design form characteristics which are included in a
generalized design mission as design parameters, mission parameters,
and mission criteria must be independent. Otherwise, a conflict
occurs and the mission cannot be satisfied. The reason is that all
design form characteristics which are included in the mission criteria
are fixed for a set of design missions, those which are included as
mission parameters are variable for a set of design missions (but
specified in a given design mission), and those which are included
as design parameters are the unspecified design characteristics in
a given set of design missions which are to be found as functions of

the mission parameters, mission criteria, and optimization criterion.

The optimization parameters are shown by Equation 5 (Chapter II) to be functions of the design parameters. Equation 1 shows that the design parameters are in turn functions of the mission parameters, mission criteria, and the optimization criterion. Therefore, a complex relationship exists between the optimization parameters and the design parameters, mission parameters, and mission criteria. This relationship is understood by studying Equations 4, 5, and 7 which show that the optimization parameters may consist of (or be functions of) certain mission parameters, mission criteria, and design parameters. Recall that the optimization criterion is used as a mapping relation in solving a generalized design mission only when one or more design parameters remain in Equation 8. Therefore, if the optimization criterion is to be used in providing mapping relations, at least one of the optimization parameters must not be expressable as a function of the mission parameters and criteria. This result provides some understanding of the relationship between the mission parameters and the optimization parameters.

It should also be noted, as pointed out by Wislicenus (6), that an operating condition in one design problem can be a design variable in another design problem. For example, the lift coefficient may be a design parameter in a design mission dealing with the entire airplane, while it could be a mission parameter in a design mission dealing with an airfoil cross-section. Generalizing, a design parameter in a particular design mission could become a mission parameter in a subdesign mission of that design mission.

Ship design example. A brief example related to ship design is presented to illustrate how mission parameters might be selected when the governing equations are difficult to establish.

Assume that the design variables in a given set of ship design problems are the total weight W, speed U, fluid density ρ, kinematic viscosity of the fluid ν, and the acceleration of gravity g. The pi theorem shows that the set W, U, ρ, ν, and g can be combined into two independent parameters. Two alternate combinations of the two parameters (as predicted from the pi theorem) are found by inspection to be

$$m_1 = \frac{Wg^2}{\rho U^6} \qquad\qquad m_2 = \frac{UW^{1/3}}{\nu g^{1/3} \rho^{1/3}} \tag{39}$$

and

$$m_1' = \frac{Wg^2}{\rho U^6} \qquad\qquad m_2' = \frac{U^3}{\nu g} \tag{40}$$

In order to determine which set is preferable, more must be known about the physical problem. Suppose that the ship drag D is to be minimized. The optimization criterion may be expressed as

$$Q = \frac{D}{W} \tag{41}$$

The drag and weight can be related by the following expressions

$$W = \rho g V \tag{42}$$

$$D = C_d \, V^{2/3} \tfrac{1}{2}\rho U^2 \tag{43}$$

where V is the volume of fluid displaced by the ship hull. Substituting Equations 42 and 43 into Equation 41 gives

$$Q = \frac{D}{W} = \frac{C_d}{2} \cdot \frac{U^2}{gV^{1/3}} = \frac{C_d}{2}\left(\frac{\rho U^6}{Wg^2}\right)^{1/3} = \frac{C_d}{2m_1^{1/3}} \tag{44}$$

where Q is seen to be a function of a type of Froude number where the length term is $V^{1/3} = (W/\rho g)^{1/3}$.

The ship design problem is solved according to Equation 44 by finding a ship hull from whose drag coefficient C_d is a minimum. It is known from theory and research experiments that C_d is a function of the hull form, standard Froude number F, and Reynolds number R_e where

$$C_d = C_d \; (F, \; R_e, \; \text{hull form}) \tag{45}$$

$$F = \frac{U}{\sqrt{g\ell}} \tag{46}$$

$$R_e = \frac{U\ell}{\nu} \tag{47}$$

Considerable experimental data has been obtained on ship hulls and tabulated as a function of F, R_e, and hull form. Consequently, the designer would want to select a pair of mission parameters such that each can be individually calculated in terms of F or R_e. According to Equations 39, 40, 42, and 46,

$$F = \frac{(V/\ell^3)^{1/6}}{m_1^{1/6}} = \frac{(V/\ell^3)^{1/6}}{(m_1')^{1/6}} \tag{48}$$

and

$$R_e = \frac{m_2}{(V/\ell^3)^{1/3}} = \frac{(m')^{1/3} m_2'}{(V/\ell^3)^{1/3}} \tag{49}$$

The preferred pair of mission parameters is therefore m_1 and m_2 rather than m_1' and m_2' since Equations 48 and 49 show that this selection provides a direct relationship with each of F and R_e. Isolating F and R_e is advantageous because each represents the influence of a different set of physical phenomena. This problem will not be carried further since the objective was to show how a preferred pair of mission parameters might be selected when the governing equations are difficult to establish.

Mission Parameter Ranges

In many cases, the objective of a design mission is a design form which will satisfy a variety of different operating conditions. For example, aircraft must operate under a variety of speed and maneuvering conditions which require large changes in wing lift coefficient. To provide this change, the wing geometry is generally made variable so that it can approximate the ideal wing required for each of the many different quasi-steady operating conditions. Similarly, an automobile must operate at a variety of speeds which requires a change in gearing in order to approximate the ideal gearing required for each speed. Alternatively, some design missions may be best satisfied by a design form which remains fixed even though it must operate under a variety of operating conditions. In such cases, the design form is compromised so that it operates adequately over the entire operating range, even though it is not the best design for any single operating situation.

To account for variable operating conditions, some or all of the mission parameters may consist of ranges of values of some parameter. For example, the mission parameter $\Delta\alpha$ could be used to represent changes in angle of attack of a wing from some average angle. Alternatively, the combination of the two parameters could be used to specify any desired range of any parameter. For example, if the take-off speed of an airplane is 100 mph and its top speed is 300 mph, the speed range could be generated by selecting one mission parameter to represent the average speed of 200 mph, and another mission parameter to represent the range of plus or minus 100 mph. An alternate method is to utilize one parameter to represent the minimum speed and another to represent the maximum speed.

The Number and Type of Design Form Solutions

By modifying the generalized design mission, both the number and type of design form solutions can be varied.

Means for reducing the number of design form solutions. If several distinct design form solutions result from each of a set of given design missions (i.e., if regions overlap), the number of distinct solutions can often be reduced by modifying the generalized design mission. The designer can do this by introducing one or more new mission parameters and mission criteria, or he can make the optimization criterion more restrictive. By using one of the three methods, the designer has a good chance of completely eliminating region overlap. However, he must have a valid reason which technically justifies introducing any change. If he cannot justify making a change, the designer should accept the region overlap, and the resulting design forms should be considered equally valid. There is no reason to believe that the overlap of regions can be prevented in all cases by any rational means.

Modification of the type of design form solutions. A generalized design mission can be changed in a variety of ways so that the type of solution is changed. One such way is to modify or eliminate a mission parameter. For example, consider the column design problem of Chapter III. If the values of E and f_p for the structural material had not been included as mission parameters, the design solutions would have been dependent upon the state of the art of structural materials since the material which had the highest values of E and f_p would have been selected. Notice that by including E and f_p as mission parameters, a unique design form results for each mission which will not change with time because it is dependent upon fixed natural phenomena. By using E and f_p as mission parameters, the designer can select various structural materials and determine how each affect the associated design forms. Consequently, care is required in setting up a generalized design mission so that the most useful type of solutions result.

82

Regions, Boundaries, and Design Form Families

These topics are related because each region of mission space corresponds to a distinct design form family, and the boundaries of each region determine its location.

Characteristics of regions. Each region is distinguished from other regions by the existence of either different physical phenomena or by a difference in the relative importance of the physical phenomena. The designer can make use of these distinguishing characteristics to determine the existence of the various regions and their location. One suggested method is to first consider all relevant physical phenomena and then consider what types of design form families might result when various combinations of these phenomena are dominant. Possible types of relevant physical phenomena might be compressive stress, buckling, bending stress, cavitation, deflection, resonance, boundary layer effects, aeroelasticity, fluid turbulence, vibration, magnetism, etc.

Natural boundaries. The reader may have noticed that some boundaries in a given mission space are caused by natural physical limits and can never change, while other boundaries may change with time because better design forms may be found. This distinction is important since it informs the designer that research or invention may help to improve the design form when the boundaries are not permanently fixed. Boundaries which are hazy and some of those which overlap may indicate that research or invention is needed.

Scaling

Scaling is defined as the process of changing the size of a design without changing its form or the associated design mission. Scaling is one of the most important aspects of engineering design. Knowledge of scaling laws permits a given design to be scaled upward or downward in size and still operate at its peak performance. The warning must be given, however, that a design form cannot merely be scaled upward or downward in size and be expected to perform well. The nondimensional design mission must also be duplicated, including the optimization criterion.

One reason why scaling is being discussed here is that, according to the definition of scaling, the scaling parameters (i.e., the parameters which must remain fixed so that a geometrically-scaled design has the same nondimensional performance characteristics) are fundamentally the same as the mission parameters and mission criteria. However, since scaling implies geometric scaling, none of the mission parameters or mission criteria which relate to the design form should be included in the set of scaling parameters because their inclusion is redundant.

Scaling parameter ranges. Scaling should not always be restricted to a single value of the scaling parameters. Often, a range in values of one of the scaling parameters can be tolerated and still provide accurate scaling. For example, if cavitation number is one of the scaling parameters, and a fully wetted hydrofoil is to be scaled, any value of the cavitation number is permissible as long as it lies above the incipient cavitation number of that hydrofoil. Consequently, the scaling laws for a given design form consist of a set of values, or a range of values, for each relevant scaling parameter. The scaling parameters are sometimes called similarity parameters or similarity relationships.

Multiple scaling parameters. Whenever more than one scaling parameter is relevant, caution is required because scaling may not be possible. An example is the scaling of ship hulls. One set of possible scaling parameters are the Froude number and the Reynolds number, Equations 46 and 47. If the fluid and g are kept constant when scaling, the Froude number requires that the ship speed change in proportion with the square root of a length dimension, while the Reynolds number requires that the speed change inversely in proportion with the length dimension. Consequently, scaling is not possible. Fortunately, exact similarity of the operating situation is not always required. In ship design, for example, it is well known that the Reynolds number has only a small effect on the performance characteristics of ships; therefore, a rather large range in Reynolds number can be tolerated.

Airplane scaling example. Scaling parameters can be developed directly from an analysis of the relevant physical phenomena using an approach similar to that described by Wislicenus (6). For example, consider the scaling of airplanes. The steady-state wing lift L is

$$L = \text{constant} \cdot C_L \ell^2 \rho U^2 \qquad (50)$$

and equals the airplane weight W which is equal to

$$W = C_s \rho_s g \ell^3 + W_o \qquad (51)$$

where ℓ is a characteristic length, C_L is the lift coefficient and is a function of wing geometry, $C_s \rho_s g \ell^3 = W_{sa}$ is the structural weight, C_s is the structural weight coefficient, ρ_s is the average mass density of the structural material, and W_o is the weight of all items except the airplane structure. The relevant scaling parameters for geometric scaling, when the lift and weight remain equal, are obtained by equating Equations 50 and 51 and nondimensionalizing them. The following groupings result

$$m_1 = \frac{\rho_s}{\rho} \cdot \frac{g\ell}{U^2} \qquad (52)$$

and

$$m_2 = \frac{W_o}{\rho \ell^2 U^2} \qquad (53)$$

Notice that C_L has been eliminated since it remains invariant for geometric scaling.

Another phenomenon which must be scaled is the structural stress caused by aerodynamic forces. The aerodynamic force is proportional to $\frac{1}{2}\rho U^2 \ell^2$ and the structural force reaction is proportional to $f\ell^2$ where f is the structural stress. Equating the two, and nondimensionalizing the equation, leads to the following grouping:

$$m_3 = \frac{\frac{1}{2}\rho U^2}{f} \qquad (54)$$

Other phenomena must be scaled, such as compressibility, maneuver-
ability, and elasticity, but these will not be considered here.

Notice that if the ratio of W_o to structural weight W_{sa} is
to remain constant (i.e., the ratio of structural weight to total
weight is invariant), Equation 53 becomes

$$m_2 = \frac{\rho_s g \ell^3}{\rho \ell^2 U^2} \cdot constant = m_1 \cdot constant$$

which shows that in this case m_2 can be eliminated since it is
essentially the same as m_1.

If an airplane is to be geometrically scaled upward, and
if W_o/W_{sa}, ρ, g, ρ_s, and f are invariant, the scaling parameters,
Equations 52 and 54, show that scaling cannot occur since the
requirements for U to change as a function of length ℓ conflict
if ℓ is changed.

If the structural material is permitted to vary, then
Equations 52 and 54 show that U^2 must vary as $\rho_s \ell$ and as f. In
other words, f/ρ_s and U^2/ρ_s must both vary as ℓ. Consequently, the
airplane can be scaled upward if f/ρ_s and U^2/ρ_s are increased in
proportion to ℓ. Unfortunately, any increase in f/ρ_s is generally
costly and difficult to achieve.

Another approach to scaling is to permit limited geometric
distortion. Suppose that the characteristic thickness b of all
structural members is permitted to increase faster than ℓ when ℓ
is increased, that the external geometric form is invariant, and that
the relative widths of all structural members vary with ℓ.

If W_o/W_{sa} remains constant, m_1 and m_3 in Equations 52 and 54 become

$$m_1' = \frac{\rho_s}{\rho} \cdot \frac{gb}{U^2} \tag{55}$$

and

$$m_3' = \frac{\frac{1}{2}\rho U^2}{f} \cdot \frac{\ell}{b} \tag{56}$$

assuming that the weight varies as $b\ell^2$, the aerodynamic force varies as ℓ^2, and the structural force varies as $b\ell$. If the structural material remains the same, Equations 55 and 56 show that U^2 must vary as b and as b/ℓ. Therefore, even this modified type of scaling cannot occur, and does not help. Now, consider an alternate type of geometric distortion (assuming again that the structural material, W_o/W_{sa}, and the external form are invariant) where b scales normally with ℓ, but where the structural material is redistributed within the wing and elsewhere so that every bit of structural material is stressed up to the allowable limit. Assuming that the airplane being scaled was not optimized in this respect, then f could be increased without changing the structural material or violating the load factors and safety requirements. Equations 52 and 54 then show that both U^2 and ℓ can be increased in proportion to f. The report by Werner (8) (which utilizes the approach to similarity developed by Wislicenus) implies that either this kind of geometric distortion, or improvements in structural material, or both, is responsible for the increase in size (and speed) of commercial propeller-driven aircraft as a function of time period.

An alternative form of geometric distortion could have been used, but it would probably have resulted in a reduction of performance. This alternative form is to reduce the lift coefficient as ℓ increases, keeping the speed constant. In this case, the external form changes without an appreciable change in the internal geometric structure. However, since C_L is reduced, the lift-to-drag ratio is reduced (assuming that the original airplane was optimized with respect to C_L), which produces a drag increase that results in a greater power requirement and fuel consumption.

As a final comment on airplane scaling, notice that the scaling parameter of Equation 52 can be placed in the following form by substituting $\ell = (W/\rho_s g)^{1/3} \cdot (\text{constant})$:

$$m_1 = \left(\frac{W^{1/2} g \rho_s^{2/3}}{\rho^{3/2} U^3}\right) \cdot \text{constant} \tag{57}$$

The parameter in the parenthesis of Equation 57 is one of the similarity parameters derived for aircraft by Wislicenus, and reported in (8). The similarity parameter of Equation 57 is therefore seen to be equivalent to that of Equation 52. Either parameter can be used for scaling the associated phenomena, the most convenient being the one which is most easily calculted from the available data.

Generalized scaling. The airplane scaling example showed that airplanes cannot be scaled geometrically when the fluid, g, and the structural material are invariant. Similar scaling difficulties will be encountered in many other kinds of scaling problems. If the desired scaling cannot be accomplished using a geometrically-scaled form, the designer should consider different kinds of geometric distortion as an alternative to the more time consuming complete redesign. Scaling where a specified geometric distortion and a specified change in the design mission is permitted is defined here as generalized scaling.

Optimized Scaling

One special type of generalized scaling is where the form of a given optimum design changes in such a manner that the resulting design is still an optimum design. This type of generalized scaling is defined here as optimized scaling.

Relationship between a set of optimized scalings and a generalized design mission. Notice that the set of design forms which results from a set of optimized scalings is the same as the family of design forms which results from a specific generalized design mission. This specific generalized design mission is the relevant one in which the only mission parameters are those which include variables that depend upon size. In other words, optimized scaling can be looked upon as the transformation from one optimized design to another where their forms belong to a special family of design forms. This special family is the one corresponding to that set of relevant design missions in which the only mission parameters are those which include variables that depend upon design size. Consequently, optimized scaling can be performed after solving the corresponding generalized design mission by merely calculating the new values of the (size-dependent) mission parameters and dimensionalizing the associated design form.

Airplane design examples. Examples of the results of optimized scaling are presented in (8) where the characteristics of
several different families of existing (and presumably optimized)
airplanes are plotted as a function of airplane weight. The forms
of any single family of (equally modern) airplanes are those forms
which would have resulted from the corresponding generalized design
mission. Some of the airplane form parameters are found to vary
while others are invariant over a relatively large weight range.
Both kinds of parameters provide valuable information which is of
future use in airplane design and research.

Optimized scaling in nature. A striking example of generalized scaling exhibited in nature which might be considered to be
optimized scaling is the scaling of mammals presented by Stahl (7).
He lists some of the form parameters of mammals which remain essentially invariant (i.e., do not vary more than about a factor of one-
half to two times an average value) over the extensive weight range
ratio of fifty million to one.[1] These results, combined with the
information on how the variable form parameters change with size,
would appear to be a significant step in the understanding of animal
physiology.

---

[1] Some of these form parameters of mammals and their average
values are: (a) heart to body weight ratio, 0.005, (b)
lung to body weight ratio, 0.011, (c) lung capacity to
blood volume ratio, 0.87, (d) heart rate to breathing rate
ratio, 4.6, (e) blood volume to total body volume ratio,
0.066, (f) blood volume to body water volume ratio, 0.10, (g)
lung/heart weight, 2.2, (h) blood weight to heart weight
ratio, 11.5. Similar relationships exist for shape and
form factors, kinematic and dynamic criteria, power and
efficiency parameters and chemical-metabolic parameters.

Application of the Design Procedure to Research

The design procedure can be applied to research either directly or by modifying it. The two kinds of procedure are described below.

Direct procedure. Since the roles of research and design are intimately related, it is sometimes difficult to determine whether a project is a design problem or a research problem. An example is the design of hydrofoil cross sections, which is presented in Appendix B, where the generalized design mission is to associate the lowest-drag hydrofoil with each of the many operating situations described by the mission parameters. For many of these operating situations, the corresponding design forms cannot be obtained directly from the results of known research. Consequently, since the design procedure permits these design forms to be specified, and since these forms are new, the generalized design problem may be considered to be a research problem as well as a design problem.

Notice that the design of hydrofoil cross sections is far removed from a complete design problem such as the design of a submarine, torpedo, or hydrofoil boat. Since the hydrofoil cross section is a subdesign problem of a propeller or strut, etc., which in turn is a subdesign problem of a complete vehicle, the design of hydrofoil cross sections is twice removed from a complete design problem. Therefore, the generalization is probably valid that the further removed a subdesign problem is from a complete design problem, the greater will be the chance that it will provide useful research results. Consequently, it is likely that many valid and important research problems can be originated and solved by studying possible subdesign or sub-subdesign problems, selecting one, and solving the resulting generalized design mission.

Another way in which the direct use of the design procedure aids in research is that it shows which regions of mission space have no known solution and may require research. Also, the procedure clearly shows the areas in which research knowledge is incomplete or inadequate.

Research procedure. The design procedure can be modified to directly solve many kinds of research problems. This modified form is called a research procedure, and appears to be a useful research technique. Occasionally, techniques similar to the kind which will be described are utilized in research, but are seldom as complete as they might be. Basically, the steps of the research procedure consist of: (a) set up a research space (i.e., a modified mission space) which is a multidimensional Cartesian space whose coordinates are nondimensional parameters called research parameters which describe a set of operating situations, (b) select a set of design forms, (c) conduct a series of tests on each form, where each test corresponds to a point in the research space, and (d) associate the resulting nondimensional performance characteristics with each point in the research space.

Examples of possible research parameters are Froude number, Reynolds number, Mach number, Webber number, pump specific speed, cavitation number, $\frac{1}{2}\rho U^2/f$, $\frac{1}{2}\rho U^2/E$, and any mission parameter used in an associated generalized design mission except those related solely to design form or to performance. The research tests consist of placing each of a set of related design forms in each operating situation described by each of various selected points in the research space. The research results are expressed in terms of values of various nondimensional performance parameters for each model. The association of the research results with points in research space is best accomplished by plotting (for each design form) the values of the various performance parameters on graphs which represent different sections of the research space. Graphs representing two-dimensional sections of research space are generally the most useful.

The research, of course, could be conducted conceptually as theoretical applied research instead of in the form of laboratory tests, if sufficient theory exists. The use of the research procedure is the same in either case.

Notice that, for a given design form, regions will appear in the research space which result from the research. These regions result from the action of different sets of physical phenomena or a difference in dominance of different physical phenomena. Examples of phenomena that might correspond to different regions are cavitation, flutter, pulsing, vibration, separation, transition, shock waves, etc.

All relevant performance parameters can be plotted on graphs of different sections of the research space. The various plotted parameters do not have to be independent; they must merely represent some kind of performance characteristic. There is no apparent reason why this procedure could not be applied to many fields cf research.

Hydrofoil example. An example of the use of the research procedure is the experimental investigation of a specific stream-lined hydrofoil model which is to be tested in a water tunnel. The research variables are the speed U, pressure P, and angle of attack α. The fixed research criteria are the water density ρ, kinematic viscosity ν, vapor pressure P_v, acceleration of gravity g, and the model characteristics. A possible set of research parameters consists of the Reynolds number R_e = Uc/ν (where c is the model chord-length), cavitation number $σ = (P-P_v)/\frac{1}{2}ρU^2$, angle of attack α, and Froude number F = U/√gc. The measured performance parameters could include the lift coefficient C_L = lift/$\frac{1}{2}ρAU^2$ (where A = plan-form area of the model), drag coefficient C_d = drag/$\frac{1}{2}ρAU^2$, static pressure coefficient C_p at various points where $C_p = (P_x-P)/\frac{1}{2}ρU^2$ (and P_x is the static pressure at the desired point), moment coefficient C_M = moment/$\frac{1}{2}ρAcU^2$, and lift-to-drag ratio L/D = C_L/C_d. All relevant physical phenomena should be observed and recorded during the experiment. The resulting performance is plotted on graphs which consist of various subspaces of research space such as σ versus α, R_e versus α, σ versus R_e, σ versus F, etc. Figure 12 illustrates one kind of graphic presentation which presents the measured lift coefficient of an imaginery, but typical, hydrofoil form as a function of σ which represents a one-dimensional section of the research space. The values of R_e and F are fixed, and α is set at whatever value is required to provide a given C_L. Also shown are the various regions with illustrations of the flow. This kind of graph provides an

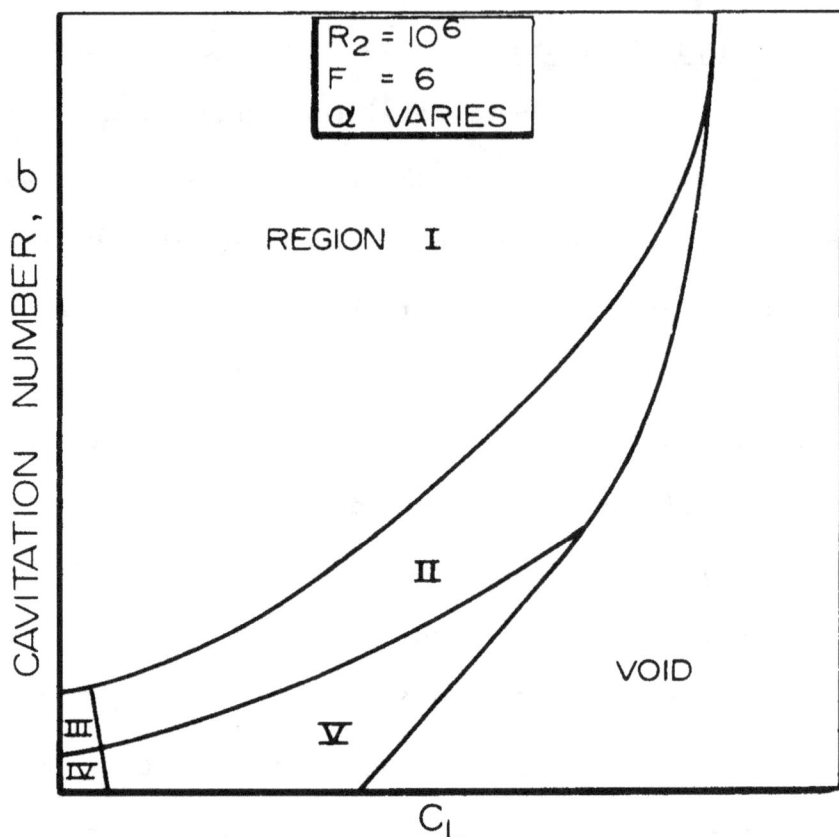

REGION

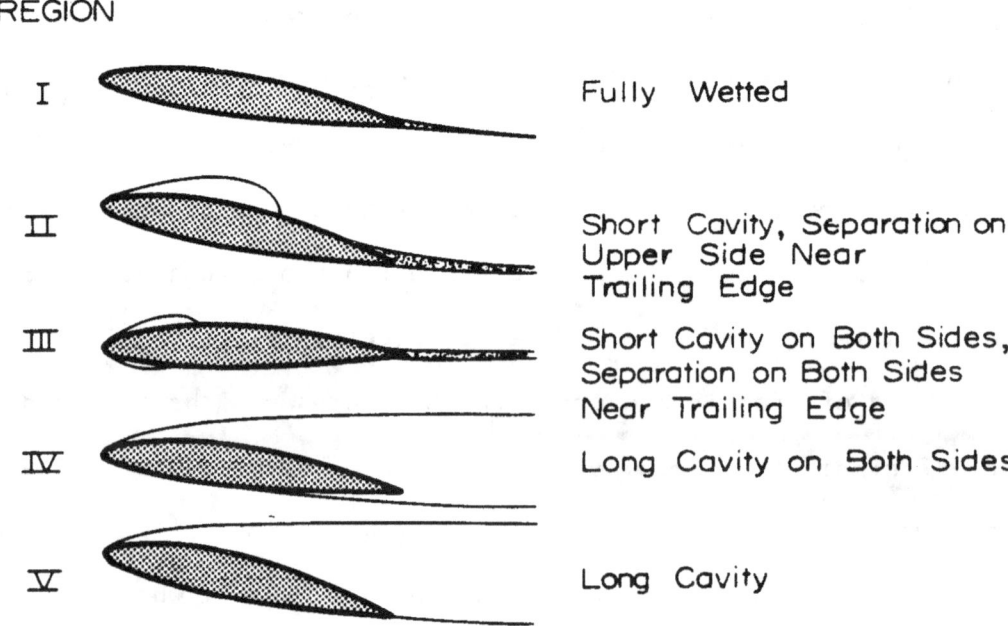

Figure 12 - Lift coefficients and flow regions of a hydrofoil
as a function of σ

excellent understanding of the effect of cavitation number on lift coefficient, and illustrates the corresponding physical phenomena.

Permissible operating range. In many engineering applications, the designer would like to know the permissible operating range of various design forms, and the factors which affect that range. Information of this kind can be obtained from research studies of the type just discussed. The results, however, are sometimes more usable if they are graphed in a different, but special, way. For example, consider the hydrofoil study which was just presented. Assume that it is desired to determine the range of σ, R_e, and F which provide a value of L/D of 15 or higher. The results of the experiment described above could be graphed in the form of Figure 13. Notice that the research parameters are inverted in order to better illustrate the permissible range. Figure 14 illustrates a required performance range for which the designer would like to find an acceptable hydrofoil. If the required performance range shown in Figure 14 lies within the operating range of a given hydrofoil form shown in Figure 13, the given hydrofoil form is acceptable.

Selection of design form families for research studies. The set of design forms which is selected for an experimental research study usually results from either a purely geometric variation or a functional variation. Either type may be useful, but the set of forms which result from a functional variation are often, but not always, the most useful type.

A set of forms represents a functional variation if a portion of the forms of the set are designed to exhibit certain desired

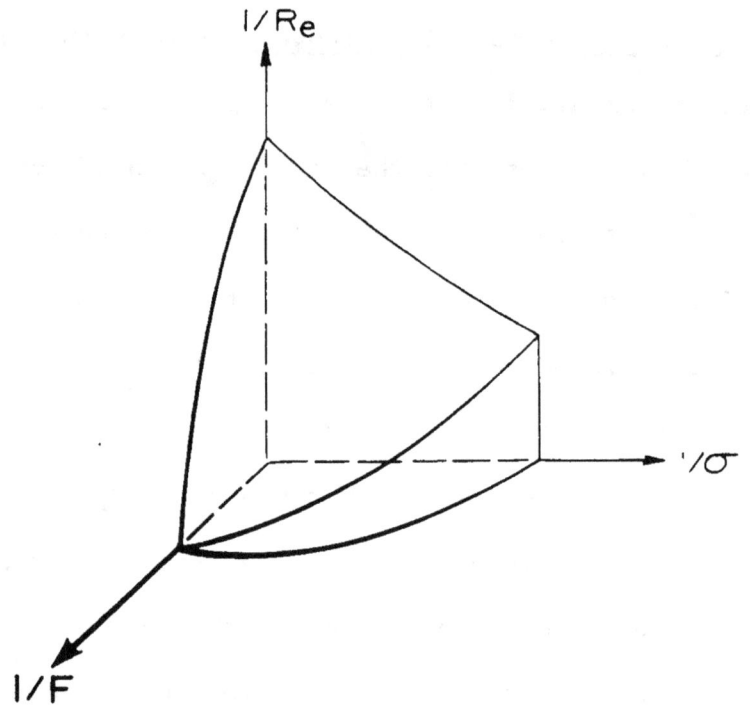

Figure 13 - Operating range of a hydrofoil

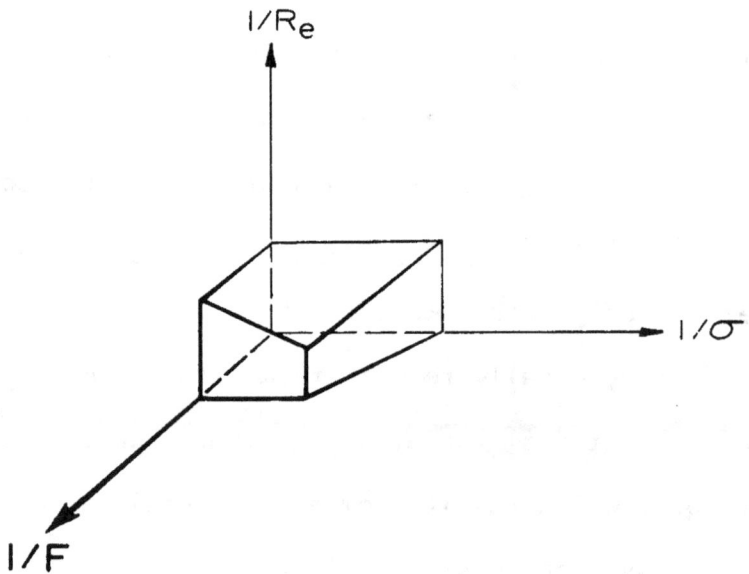

Figure 14 - Required hydrofoil operating range

performance characteristics, and the rest are designed to vary a small amount from them. The experimental objective is to determine whether the desired result is obtained, and if not, whether any of the models designed for a small variation from the desired objective are more suitable.

An example of an experimental research study based upon a functional variation is one in which a series of hydrofoil forms are tested to determine which of several low-drag forms has the best cavitation resistance. All forms could have the same strength or thickness-to-chord ratio, but different calculated pressure distributions in ideal fluid flow, all of which are fairly uniform, except near the trailing edge. The angle of attack could be fixed at the ideal angle. An alternate series of tests might be conducted which correspond to a specific angle of attack range, such as $\pm$ 2.0 degrees from the ideal angle of attack.

Future Engineering Design Theories

When a field is in the state of rapid evolution, as many are at this time, including engineering design theory, it is difficult to predict future developments. Although engineering design theory may follow a variety of different paths, it is likely that at least one path will utilize a rigorous mathematical approach.

Applied group theory. One possible approach to a rigorous mathematical foundation for a design theory is the utilization of group theory, which was suggested by Dr. D. P. Hoult in a graduate course which he taught.[1] A group is a mathematical term defined as a set of elements together with an associative binary operation defined on the set such that: (a) an identity element exists in the set, and (b) the inverse of each element exists in the set. By looking upon the set of transformations from one design fcrm to another as elements of a group, Hoult suggests in the course notes that a group can be defined. He briefly suggested possible ways in which the properties of mathematical groups might be helpful in solving a design problem. Considerable work must be done, however, to develop these ideas, or others, to the point where they are clearly shown to contribute meaningfully to the solution of a design problem. Appendix A shows how the transformations from one design form to another can be looked upon as elements of a group.

Another approach to a more rigorous mathematical foundation was recently presented by Dr. J. W. Bond after reviewing this generalized design procedure.[2] He suggests a mathematical framework in which the different aspects of this design theory may be treated. In particular, he treats the problem of how to select the various nondimensional parameters and relates them to generalized scaling.

---

[1]   A brief outline of a possible approach to a theory of design using group theory was included in a set of course notes for a course taught by D. P. Hoult at The Pennsylvania State University in June, 1967, called "Applied Group Tleory" and listed as AROE 520.

[2]   An informal technical memorandum on design theory by Dr. J. W. Bond will be filed with the Department of Aerospace Engineering so that all interested readers may receive a copy upon request.

Design theory relationships. Perhaps some of the design relationships developed herein could aid in developing an improved engineering design theory. The following is a list of some of these relationships: (a) a generalized design mission is uniquely determined by a design objective, a set of mission criteria, a set of mission parameters, and an optimization criterion; (b) a variety of generalized design missions can be derived from a given design mission; (c) the number and nature of relevant (dimensional) variables which describe a generalized mission is unique; (d) the set of mission parameters is not unique, but one set may be preferable; (e) a different family of design forms corresponds to each region of mission space; (f) a single design form is associated with each point in a region of mission space; (g) a different set of mapping relations corresponds to each region of mission space; (h) a different set of physical phenomena or a different relative dominance within a set of physical phenomena corresponds to each region of mission space, to each family of design forms, and to each set of mapping relations; (i) the number of regions in mission space is equal to the number of different sets of mapping relations; (j) the number of mapping relations corresponding to each region is equal to the number of design form parameters; (k) the number of mapping relations which must be obtained from the optimization criterion is equal to the number of design form parameters minus the number of design equations; (l) if two regions overlap in mission space, two different design forms will correspond to each

point in the overlapping space; and (m) if the design forms corresponding to mission space vary continuously across a boundary, the
two design form families corresponding to the adjacent regions have
one subfamily of design forms in common.

CHAPTER V

DESIGN OF SUBMERGED VEHICLES

Submerged vehicles are defined as the class of self-propelled devices which travel through a fluid and whose weight is principally supported by buoyancy force. The treatment of submerged vehicles in this chapter is as general as possible, and will apply to submarines, torpedoes, and airships. The effect of technological advancements on the design form will also be presented.

Generalized Mission

The objective of the generalized mission is to determine the form of a submerged vehicle and the relative size of its major components. The mission criteria are: (a) steady state operation, (b) horizontal travel, (c) fully-wetted vehicle, (d) turbulent boundary layer, and (e) no free-surface effects exist. The optimization criterion requires that the size of the vehicle is to be minimized. The mission variables are listed below.

Mission variables. The selected set of mission variables consists of the speed U; range R; the volume of vehicle components V_o which are independent of speed, range, or density requirements; the average mass density ρ_o of these components; acceleration of gravity g; density of the fluid ρ; kinematic viscosity of the fluid ν; maximum operating depth z; static pressure P at the minimum operating depth; vapor pressure P_v of the fluid at the minimum operating depth; average volume per unit of net power output α_p, and average mass density ρ_p, of all components which vary with the power output; average volume per unit of net energy output α_e, and average mass density ρ_e, of all components which vary with the energy output; average mass density of the buoyancy source ρ_b; and average mass density of the vehicle ρ_v. Summarizing, the sixteen mission variables are U, R, V_o, ρ_o, g, ρ, ν, z, P, P_v, α_p, ρ_p, α_e, ρ_e, ρ_b, and ρ_v.[1]

The mission variables listed above were selected because it is assumed that these items would be specified in a typical submerged vehicle design problem. Other sets of mission variables could have been selected. To answer questions which the reader may have at this point, the reasons for the selection of these variables are presented next.

Reasons for selecting the mission variables. Generally speaking, the speed, range, and maximum depth (of underwater vehicles) are specified. The minimum operating depth, or the corresponding static pressure P, is often important for underwater vehicles because this is the depth where cavitation is most likely to occur Cavitation is to be avoided because the mission criteria contain the specification that the vehicle is to be fully wetted. The fluid through which the vehicle travels is usually specified. The relevant

---

[1] In the case of airship design, P_v and the operating depth z are not relevant. Without conflict of nomenclature, P can be defined as the static pressure of the air for the special case of airship design.

fluid characteristics for vehicle design are ρ, ν, and P_v.

The influence of gravity appears to be relevant since the vehicle weight is important when the net vehicle density ρ_v is specified, so g is included. To control the net density of a vehicle, a buoyancy section or purposefully loose packing of certain vehicle components is often required. The effective net density of the buoyancy source is ρ_b.

In vehicle design problems, the payload density and volume are generally given, or can be calculated. Also, the density and volumes of the electronic components, flooded sections and spaces, controls, and stabilizing surfaces can be approximately determined before the design of the complete vehicle is started. Since the maximum operating depth is known, the volume and density of all structural parts associated with the payload and the other items just mentioned can be estimated as a subdesign problem which is conducted before beginning to solve the generalized design mission. Notice that the payload and other items do not always have to be placed in a pressure resistant hull. Whatever structure is to be utilized, the net density ρ_o and volume V_o of these components and their associated structure can be closely estimated before beginning the generalized design mission.

Reasons have been given for the selection of all variables except α_p, ρ_p, α_e, and ρ_e. Inclusion of these variables as mission variables means that the power plant, propulsor, and fuel source are known at the beginning of the generalized design mission. This assumption is justified by the following considerations:

(a) practical vehicle designs are based upon a variety of factors, such as reliability, simplicity, safety, ease of maintenance and repair, efficiency, etc., and these factors have not been included as specifications for the selected set of design problems· and (b) the generalized design procedure represents an infinite number of design missions; therefore, the designer can select several specific design missions (which correspond to different values of α_p, ρ_p, α_e, ρ_e) in order to determine the effect on design form of utilizing the different combinations of power plant, propulsor, and fuel source which he thinks should be considered.

A final comment on the selection of mission variables is that for any given design mission, it is assumed that several sub-design problems have already been solved so that the required values of the different mission variables can be determined. In other words, it is assumed that the payload, electronics, controls, power plant, propulsor, fuel source, buoyancy source, etc., have been optimally designed by taking into account all relevant practical factors. Since the vehicle speed, maximum depth, and fluid characteristics are known, each of the variables V_o, ρ_o, α_p, ρ_p, α_e, ρ_e, and ρ_b are assumed to include all effects of depth, speed, and the fluid. Also, all vehicle components are assumed to include their respective portion of the vehicle structure. For instance, V_o, α_p, and α_e should include the volume of any associated structure, including a portion of the vehicle pressure hull, if a pressure hull exists. The above assumptions are necessary because no rational method exists for determining their values in a generalized design mission. In other words, the selection of specific values for each of the mission variables depends so much on the nature of each specific design mission that no way exists for assigning them a generalized value.[1]

Possible mission parameters. Since sixteen mission variables were selected, and all are dimensional, the pi theorem predicts thirteen independent nondimensional parameters. One possible set is $R/V_o^{1/3}$, $U^2/gV_o^{1/3}$, $UV_o^{1/3}/\nu$, $z/V_o^{1/3}$, $P/\frac{1}{2}\rho U^2$, $P_v/\frac{1}{2}\rho U^2$, ρ_o/ρ, ρ_v/ρ, ρ_p/ρ, ρ_e/ρ, ρ_b/ρ, $\alpha_p\rho g U$, and $\alpha_e\rho g R$.

The set of nondimensional parameters is now briefly examined to determine if all parameters are relevant. First, consider the parameter $z/V_o^{1/3}$ which represents the maximum depth. This parameter can be eliminated as a mission parameter because z is utilized in a subdesign problem before the design mission is set up in order to determine the depth effect on the hull design, etc., of V_o, ρ_o, α_p, ρ_p, α_e, ρ_e, and ρ_b; consequently, z is not relevant as a separate variable since it has no further influence on the design mission.

Another change in the list of variables is that $P/\frac{1}{2}\rho U^2$ and $P_v/\frac{1}{2}\rho U^2$ can be combined into a new parameter $(P-P_v)/\frac{1}{2}\rho U^2$ which is

---

[1] Exact values for all selected variables cannot always be calculated at the beginning of a design problem, so approximate values must be estimated and later refined after obtaining a preliminary solution to the design problem. The problem should then be reworked if the estimated values are found to be significantly in error.

known as the cavitation number σ. The individual parameters $P/\frac{1}{2}\rho U^2$ and $P_v/\frac{1}{2}\rho U^2$ are not relevant to the generalized design mission.[1]

The new set of eleven candidates for mission parameters is: $R/V_o^{1/3}$, $U^2/gV_o^{1/3}$, $UV_o^{1/3}/\nu$, $(P-P_v)/\frac{1}{2}\rho U^2$, $\alpha_p\rho gU$, $\alpha_e\rho gR$, ρ_v/ρ, ρ_o/ρ, ρ_b/ρ, ρ_p/ρ, and ρ_e/ρ.

Possible Design Forms

Figure 15 illustrates the variety of vehicle forms which might be associated with the different design missions represented by points in mission space.

A problem arises now, because how can a point in the mission space defined by the eleven parameters listed above determine which of the vehicle forms should be selected? Apparently, a new mission parameter is needed which describes the type of vehicle desired. Let the symbol for the new mission parameter be T, where T represents the type of vehicle.

Physical Relationships

The optimization goal is considered first. Since this goal is to minimize the vehicle volume, the first physical relationship should be an expression for the vehicle volume V, which may be

---

[1] Although the variables P and $\frac{1}{2}\rho U^2$ are used in the case of airship design to determine the structure of the buoyancy chamber, their effect has already been included in the variable ρ_b; consequently, $P/\frac{1}{2}\rho U^2$ is not relevant, even in the case of airship design.

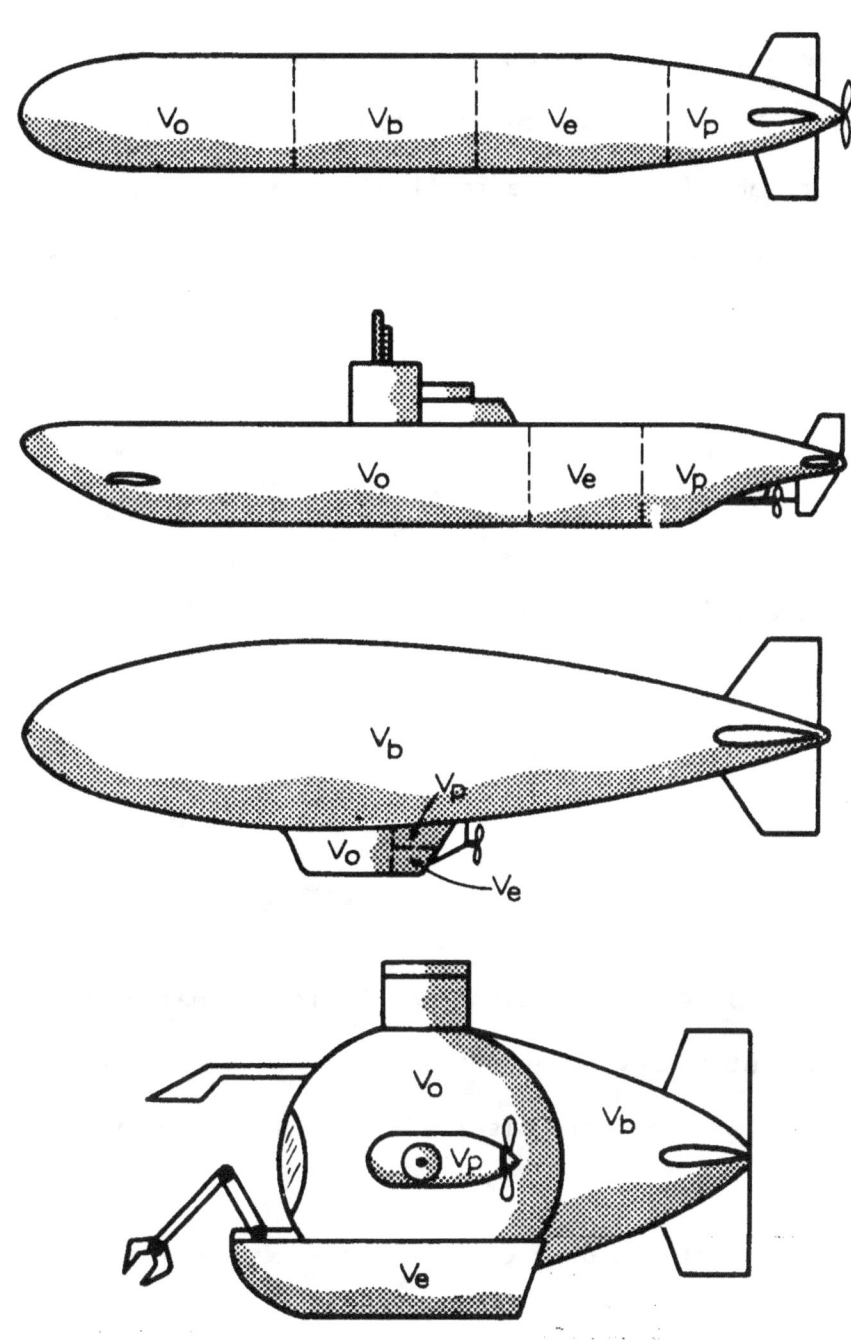

Figure 15 - Possible forms of submerged vehicles

108

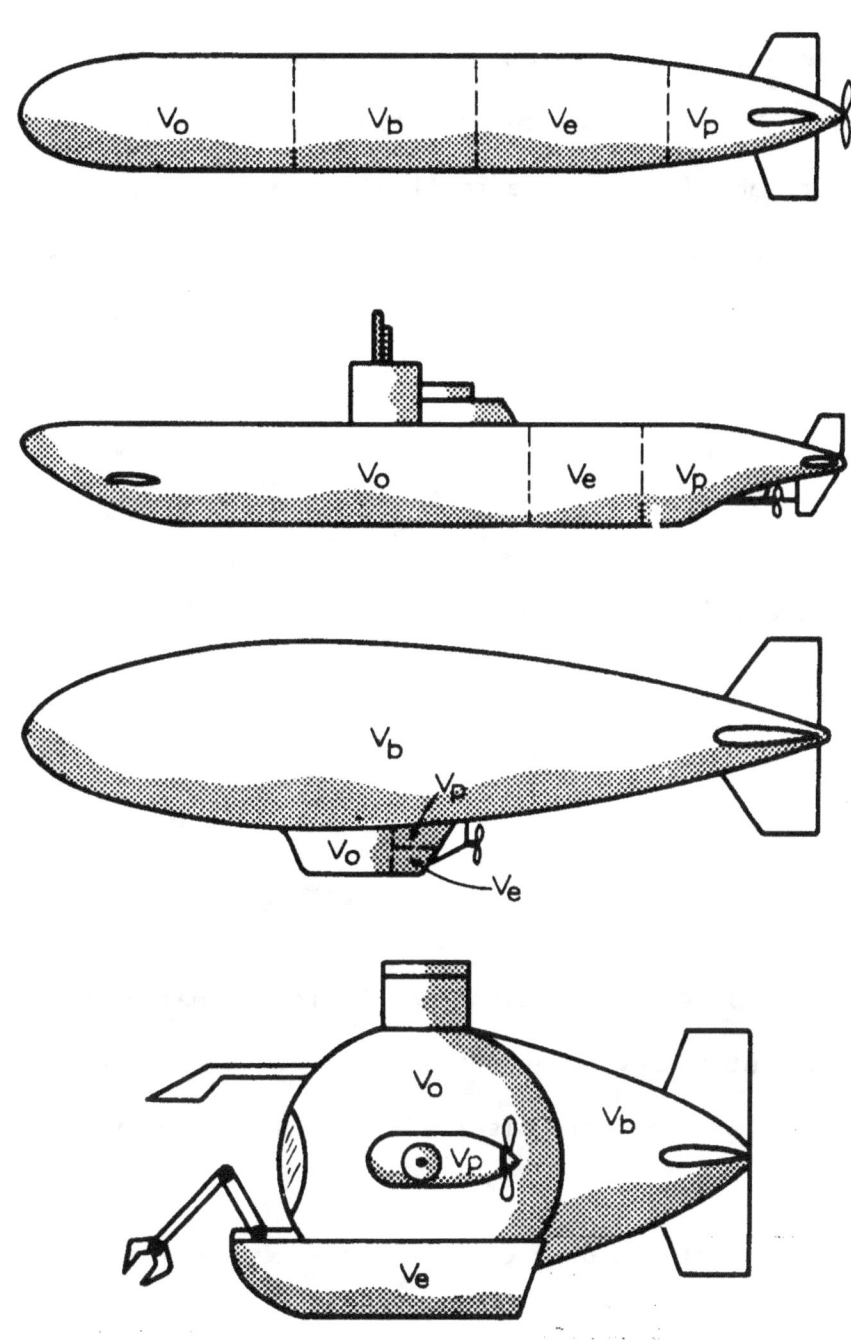

Figure 15 - Possible forms of submerged vehicles

expressed as

$$V = V_o + V_p + V_e + V_b \tag{58}$$

where V_p is the volume of components which vary with the power output, V_e is the volume of components which vary with the energy output, and V_b is the (effective) volume of the buoyancy source, if any. As mentioned earlier, the associated structure is included in each volume component. The optimization criterion Q is the nondimensional vehicle volume, or

$$Q = \frac{V}{V_o} = 1 + \frac{V_p}{V_o} + \frac{V_e}{V_o} + \frac{V_b}{V_o} \tag{59}$$

where Q is to be minimized. Utilizing the definitions of α_p and α_e,

$$V_p = \alpha_p DU \tag{60}$$

$$V_e = \alpha_e DR \tag{61}$$

where D is the vehicle drag, DU is the net power output, and DR is the net energy output of a vehicle. Notice that the power plant efficiency and propulsor efficiency are included in the values of both α_p and α_e.

Let the drag coefficient C_d be defined by the following expression:

$$D = C_d V^{2/3} \tfrac{1}{2} \rho U^2 \tag{62}$$

where C_d is known to be a function of the Reynolds number R_e and the vehicle form where

$$R_e = \frac{U\ell}{\nu} \tag{63}$$

ℓ is the length of the vehicle, and

$$C_d = C_d (R_e, \text{ vehicle form}) \tag{64}$$

Substituting Equations 60 to 62 into Equation 59 yields

$$Q = \frac{V}{V_o} = 1 + C_d \frac{V^{2/3}}{V_o} \tfrac{1}{2}\rho U^2 (\alpha_p U + \alpha_e R) + \frac{V_b}{V_o} \tag{65}$$

Consequently, Q can be minimized only by minimizing both C_d and V_b, since all of the other variables are mission variables and are therefore fixed for any specific design mission.

The physical relationship for vehicle weight W is

$$W = W_o + W_p + W_e + W_b \tag{66}$$

By substituting g times the relevant density and volume for the weight of each component and dividing by g, Equation 66 becomes

$$\frac{W}{g} = \rho_v V = \rho_o V_o + \rho_p V_p + \rho_e V_e + \rho_b V_b \tag{67}$$

No other physical relationships appear to be relevant at this point, so the equations are placed in nondimensional form and rewritten to obtain nondimensional groupings of variables. Multiplying Equation 65 by $(V_o/V)^{2/3}$, and keeping in mind the set of possible mission parameters, yields

$$\left(\frac{V}{V_o}\right)^{1/3} = \left(\frac{V_o}{V}\right)^{2/3} + \frac{C_d}{2}\left(\frac{\alpha_p \rho U^3}{V_o^{1/3}}\right) + \frac{C_d}{2}(\alpha_e \rho U^2)\left(\frac{R}{V_o^{1/3}}\right) + \frac{V_b}{V_o}\left(\frac{V_o}{V}\right)^{2/3} \tag{68}$$

where the unknowns are V_o/V, C_d, and V_b/V_o. The latter term can be obtained by first rewriting Equation 67 as

$$\frac{V_b}{V_o} = \frac{\rho_v}{\rho_b}\frac{V}{V_o} - \frac{\rho_o}{\rho_b} - \frac{\rho_p}{\rho_b}\frac{V_p}{V_o} - \frac{\rho_e}{\rho_b}\frac{V_e}{V_o} \tag{69}$$

By substituting Equations 60 to 62, Equation 69 becomes

$$\frac{V_b}{V_o} = \frac{\rho_v}{\rho_b}\frac{V}{V_o} - \frac{\rho_o}{\rho_b} - \frac{\rho_p}{\rho_b}\frac{C_d}{2}\left(\frac{V}{V_o}\right)^{2/3}\left(\frac{\alpha_p \rho U^3}{V_o^{1/3}}\right) - \frac{\rho_e}{\rho_b}\frac{C_d}{2}\left(\frac{V}{V_o}\right)^{2/3}(\alpha_e \rho U^2)\left(\frac{R}{V_o^{1/3}}\right) \tag{70}$$

where the unknowns are V_b/V_o, V_o/V, and C_d. Substituting Equation 70 into Equation 68 and rewriting, yields

$$\left(\frac{V}{V_o}\right)^{1/3} - \left(\frac{V_o}{V}\right)^{2/3}\left(\frac{\rho_o-\rho_b}{\rho_v-\rho_b}\right) = \frac{C_d}{2}\left(\frac{\alpha_p \rho U^3}{V_o^{1/3}}\right)\left(\frac{\rho_p-\rho_b}{\rho_v-\rho_b}\right) + \frac{C_d}{2}(\alpha_e \rho U^2)\left(\frac{R}{V_o^{1/3}}\right)\left(\frac{\rho_e-\rho_b}{\rho_v-\rho_b}\right) \tag{71}$$

An alternate form of Equation 71 which pertains to the case when vehicle density is unimportant, is Equation 68 less the last term, or

$$\left(\frac{V}{V_o}\right)^{1/3} - \left(\frac{V_o}{V}\right)^{2/3} = \frac{C_d}{2}\left(\frac{\alpha_p \rho U^3}{V_o^{1/3}}\right) + \frac{C_d}{2}(\alpha_e \rho U^2)\left(\frac{R}{V_o^{1/3}}\right) \tag{72}$$

If C_d is known, Equation 71 and the three following equations can be solved for the case when density is important to provide the four form parameters V_o/V, V_p/V, V_e/V, and V_b/V:

$$\frac{V_p}{V} = \frac{C_d}{2} \left(\frac{V_o}{V}\right)^{1/3} \left(\frac{\alpha_p \rho U^3}{V_o^{1/3}}\right) \tag{73}$$

$$\frac{V_e}{V} = \frac{C_d}{2} \left(\frac{V_o}{V}\right)^{1/3} (\alpha_e \rho U^2) \left(\frac{R}{V_o^{1/3}}\right) \tag{74}$$

$$1 = \frac{V_o}{V} + \frac{V_p}{V} + \frac{V_e}{V} + \frac{V_b}{V} \tag{75}$$

where Equations 73 and 74 were obtained from Equations 60 to 62, and Equation 75 resulted from Equation 58. The design equations for the case when density is not critical are Equations 72 to 74. Notice that for either case, the number of design equations is equal to the number of design form parameters, so the expression for Q is not needed to provide further equations. (Recall that Q was used earlier to show that C_d must be minimized.)

Evaluation of C_d. The value of C_d cannot be obtained directly in view of the following facts: (a) C_d is a function of R_e and the design form, (b) the design form is known only after the design equations are solved, (c) either the value of C_d or an equation for C_d as a function of the design form is needed in order to solve the design equations, and (d) the relationship of C_d to design form is far too complicated to be expressed in equation form.

Consequently, the best method for solving the design equations appears to be an iterative process consisting of the following steps: (a) select a preliminary value of C_d, (b) solve the design equations for the design form, (c) utilize the solution to calculate an improved value of C_d, and (d) repeat Steps (b) and (c). The value of C_d should converge quickly; in fact, a repeat solution of the design equations may not be needed if the subdesign problem outlined in the next section is carried out.

Determination of vehicle form. The designer can determine the approximate form of the vehicle by solving a subdesign mission in which the mission criteria are the cavitation number σ, the parameter $UV_o^{1/3}/\nu$, and the vehicle type T. Knowledge of the value of the mission parameter $UV_o^{1/3}/\nu$ provides a rough value of the Reynolds number since the parameter can be written as

$$\frac{UV_o^{1/3}}{\nu} = \left(\frac{U\ell}{\nu}\right)\left(\frac{V_o^{1/3}}{\ell}\right) \tag{76}$$

The value of $V_o^{1/3}/\ell$ can be approximated from a preliminary form study in which the designer guesses at the relative sizes of all components, sketches a possible design form, and then estimates the length of the vehicle from the known value of V_o.

The cavitation number σ may or may not have an effect on the design form solutions of the generalized design mission. If the value of σ is low (i.e., below 1.0 to 2.0), σ may affect the design forms by requiring them to be more highly streamlined; also, a special type of propulsor may be required to avoid cavitation.

Consequently, during the preliminary study on design form, the designer should include the effect of σ, if it is found necessary to do so. At very low values of σ, of around 0.1 to 0.3, the vehicle shape must be long and slim in order to avoid cavitation; also, one of the less common types of propulsors is generally required to avoid cavitation.

The vehicle body form which has the best cavitation resistance consists of a small tailcone attached to the rear of a uniform-pressure body shape developed by Munzner and Reichardt (15), which is described by the equation

$$\left(\frac{x-\ell/2}{\ell/2}\right)^2 + \left(\frac{y}{d/2}\right)^{2.4} = 1 \tag{77}$$

where ℓ is the body length, d is the maximum diameter, x is the longitudinal distance from the nose, and y is the radius. The relationship of ℓ and d to cavitation number σ, for small values of σ, is

$$\frac{d}{\ell} = 0.793 \, \sigma^{5/8} \tag{78}$$

In the cases where the cavitation number is not highly critical, and a uniform-pressure body is not needed, a variety of forms can be utilized such as a circular cylinder with a rounded nose and a streamlined tail. If the cavitation number is in the moderate-to-low range, the design of the nose form of a cylindrically-shaped vehicle will probably be affected by σ. Rouse and McNown (16) present valuable data on the shape of nose sections as a function of σ.

In an analysis of the drag of a wide variety of streamlined torpedo-like vehicles, Brooks and Lang (17) show that the value of C_d, as defined by Equation 62, varies somewhat with R_e, but only slightly with design form over a wide range of length-to-diameter ratios, as long as the boundary layer is turbulent, the nose is reasonably rounded, and the tailcone has a length-to-diameter ratio of about 2.5 or more. A typical value of C_d is 0.022. For high-drag vehicles like most of the small research submarines which have many kinds of appendages and poorly streamlined shapes, the value of C_d could be several-fold larger, but can be quickly calculated using information from Hoerner (18); the value of C_d in this case will vary only slightly with either R_e or with small to moderate changes in form, if the degree of streamlining remains the same. Assuming that the designer takes care in streamlining the vehicle form as much as possible, he will arrive at a value for C_d which should be close to a minimum value.

Selection of the Mission Parameters

Before selecting the mission parameters, consider the nature of the parameter C_d. Notice that it is evaluated before analysis of the generalized design mission is begun, just like any of the possible mission parameters. In fact, C_d has all of the properties of a mission parameter, and should therefore be added to the list of possible mission parameters.

After studying the design equations, Equations 71 to 75, and the list of possible candidates, the best set of mission parameters appears to be the following:

$$m_1 = \frac{C_d \alpha_p \rho U^3}{2\, V_o^{1/3}} \tag{79}$$

$$m_2 = \frac{C_d \alpha_e \rho U^2 R}{2\, V_o^{1/3}} \tag{80}$$

$$m_3 = \frac{C_d \alpha_p \rho U^3}{2\, V_o^{1/3}} \left(\frac{\rho_p - \rho_b}{\rho_v - \rho_b}\right) \tag{81}$$

$$m_4 = \frac{C_d \alpha_e \rho U^2 R}{2\, V_o^{1/3}} \left(\frac{\rho_e - \rho_b}{\rho_v - \rho_b}\right) \tag{82}$$

$$m_5 = \frac{\rho_o - \rho_b}{\rho_v - \rho_b} \tag{83}$$

assuming that C_d has been approximately evaluated from the following function:

$$C_d \doteq C_d \left(\frac{U V_o^{1/3}}{\nu},\ \sigma,\ T\right) \tag{84}$$

Therefore, the parameters $U V_o^{1/3}/\nu$, σ, and T have already been used for this problem and do not appear in the above set of mission parameters. Notice that the familiar Froude number is hidden in both m_1 and m_2 since

$$m_1 = \frac{C_d}{2} \cdot \frac{U^2}{gV_o^{1/3}} \cdot \alpha_p \rho g U$$

$$m_2 = \frac{C_d}{2} \cdot \frac{U^2}{gV_o^{1/3}} \cdot \alpha_e \rho g R$$

Also, notice that m_3 could have been set equal to $(\rho_p - \rho_b)/(\rho_v - \rho_b)$, but it was selected as presented because it appears in this form in Equation 71 and will be seen to be used in this form later in graphing mission space. A similar remark holds for m_4.

Status of the Problem

By using a subdesign problem to calculate C_d, the generalized design mission has become considerably simplified. The mission parameters are m_1 to m_5. The design form parameters are V_o/V, V_p/V, V_e/V, and V_b/V. The design equations are Equations 71 to 75. The optimization criterion is not needed at this stage of the problem. All five mission parameters are required when vehicle density is important; otherwise, only m_1 and m_2 are relevant. No boundaries will appear in mission space since no critical values are apparent and the same set of mission relations are relevant everywhere in the space. The shape of the vehicle may be significantly affected by σ if σ is below about 1.0 or 2.0.

Power- and Energy-Limited Vehicles

The simplest set of design missions is the generalized mission described by mission parameters m_1 and m_2 when vehicle density is not important. The design forms, for a given value of V_o, are then dependent only on power and energy requirements. The mapping relations, Equation 72 to 74, when placed in terms of m_1 and m_2, become

$$\left(\frac{V}{V_o}\right)^{1/3} - \left(\frac{V_o}{V}\right)^{2/3} = m_1 + m_2 \tag{85}$$

$$\frac{V_p}{V} = \left(\frac{V_o}{V}\right)^{1/3} m_1 \tag{86}$$

$$\frac{V_e}{V} = \left(\frac{V_o}{V}\right)^{1/3} m_2 \tag{87}$$

The point $m_1 = m_2 = 0$ in the (m_1, m_2) space maps into $V_o/V = 1.0$, meaning that the entire vehicle consists of V_o. This type of design form corresponds to the limiting case of a low-speed, short-range form from the viewpoint of minimum size.

The mapping of either the line $m_1 = 0$ or $m_2 = 0$ is illustrated in Figure 16 where V_o/V is graphed as a function of m_1 or m_2. The ratio V_p/V or V_e/V is simply $1 - V_o/V$, as seen by Equation 75 when $V_b/V = 0$.

The mapping result of the entire (m_1, m_2) space is illustrated by Figure 17 where V_o/V, V_p/V, and V_e/V are each graphed as a function of m_1 and m_2.

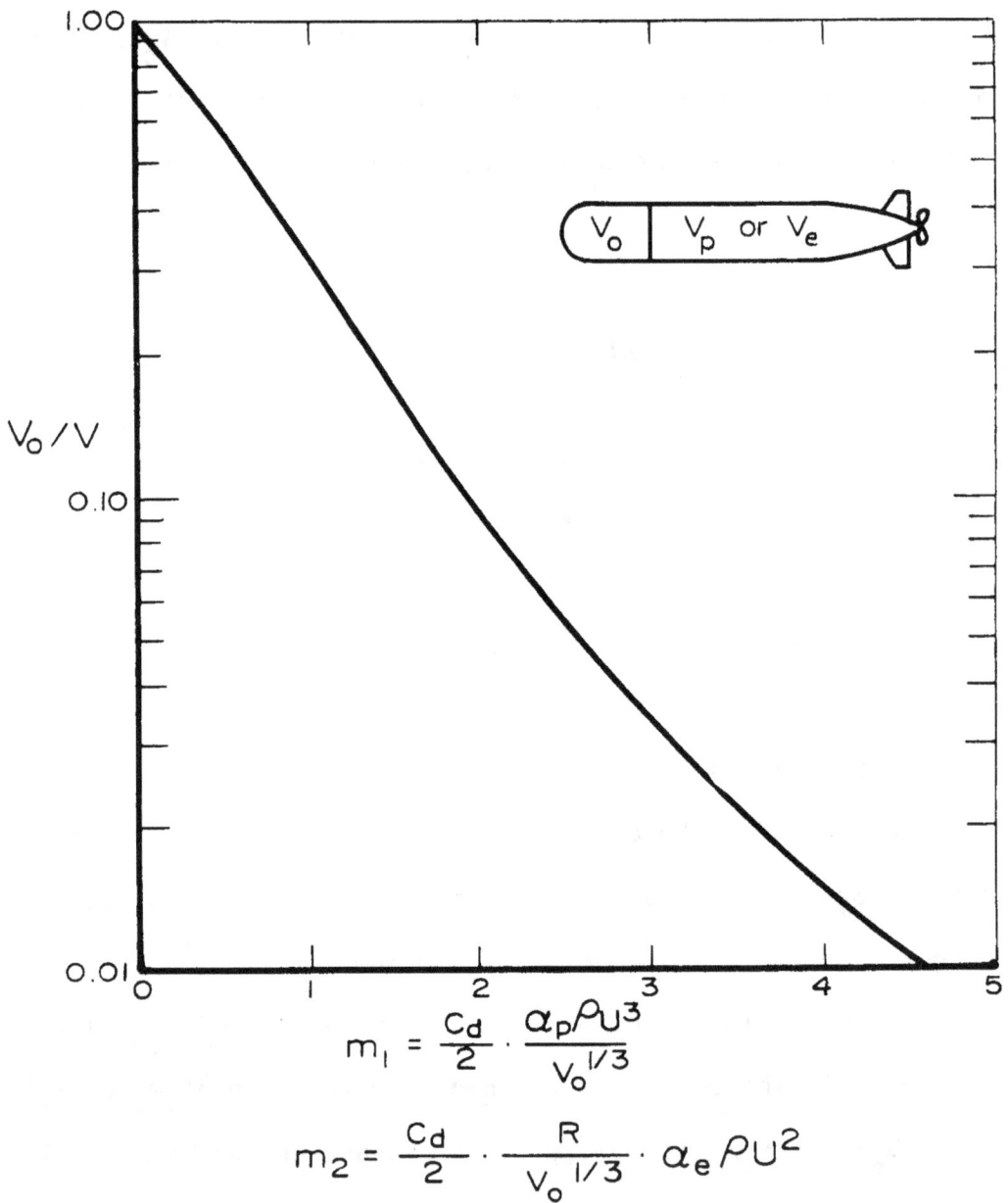

Figure 16 — Vehicle limited by power or energy

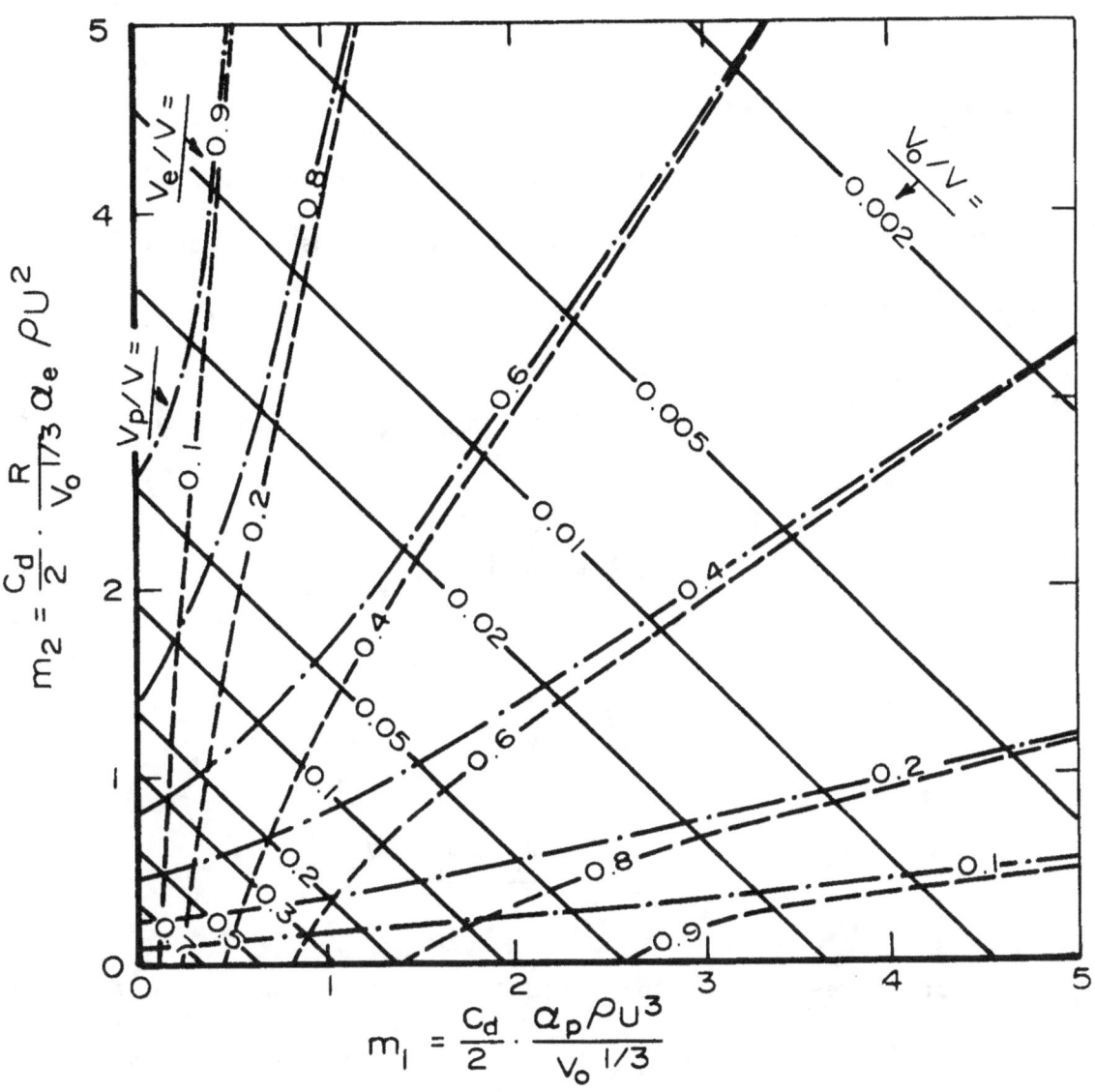

Figure 17 - Vehicle limited by power and energy

The mapping for the special case when vehicle volume is limited, is shown in Figure 18 to illustrate how easily the generalized design mission can be turned into a size-limited design mission where a critical boundary line now exists. The example of Figure 18 corresponds to the set of missions where $V \leq 20V_o$.

Power-, Energy-, and Density-Limited Vehicles

When vehicle density is important, the required mission parameters are m_1 to m_5. The mapping relations are Equations 71, 73, 74, and 75 which, after substituting m_1 to m_5, become

$$\left(\frac{V}{V_o}\right)^{1/3} - \left(\frac{V_o}{V}\right)^{2/3} m_5 = m_3 + m_4 \tag{88}$$

$$\frac{V_p}{V} = \left(\frac{V_o}{V}\right)^{1/3} m_1 \tag{89}$$

$$\frac{V_e}{V} = \left(\frac{V_o}{V}\right)^{1/3} m_2 \tag{90}$$

$$\frac{V_b}{V} = 1 - \frac{V_o}{V} - \frac{V_p}{V} - \frac{V_e}{V} \tag{91}$$

The mapping result is illustrated in Figure 19 where V_o/V is graphed as a function of m_3 and m_4 for three different values of m_5. The values of V_p/V, V_e/V, and V_b/V can be obtained from Equations 89 to 91.

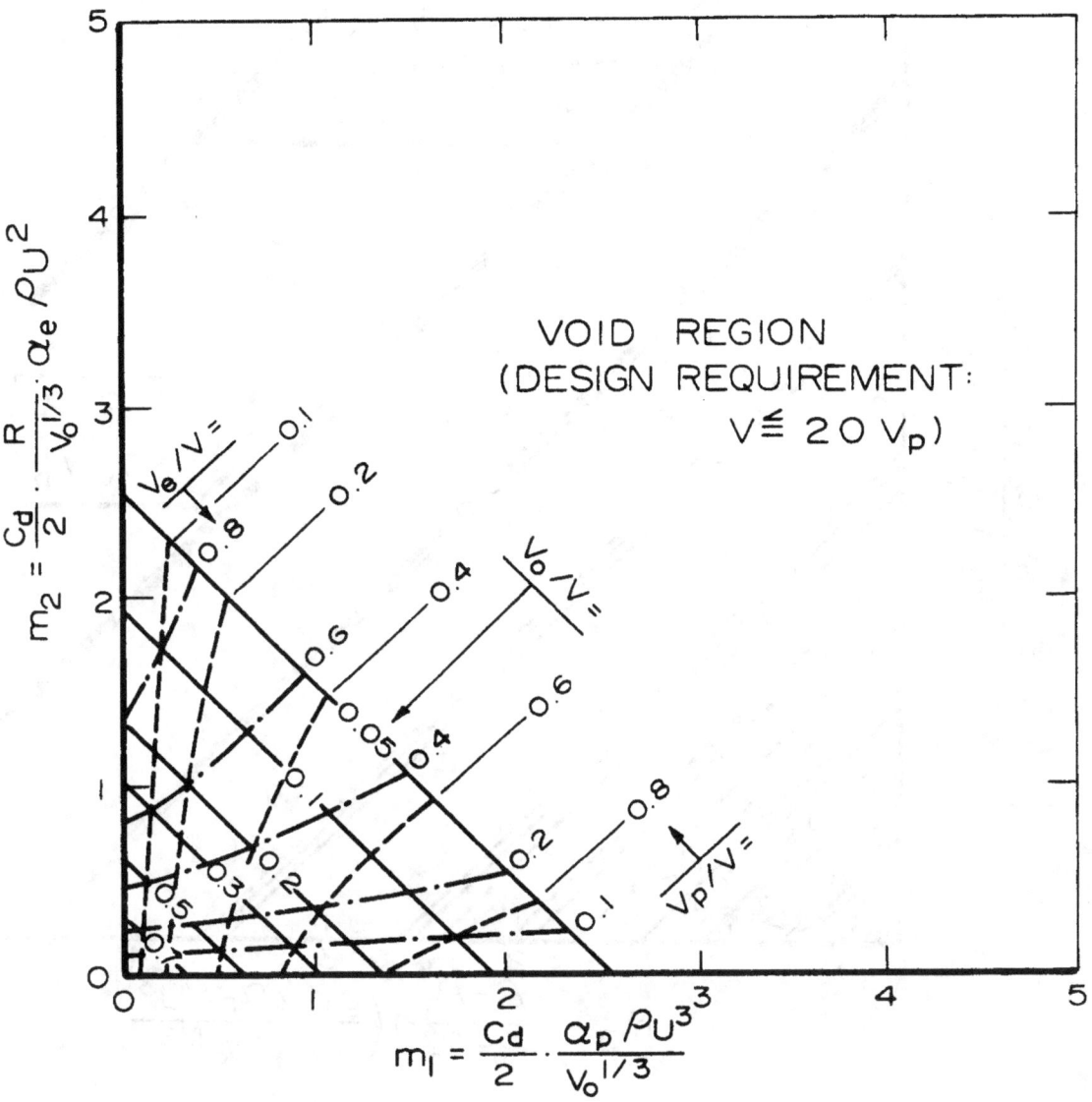

Figure 18 - Vehicle limited by power, energy, and volume

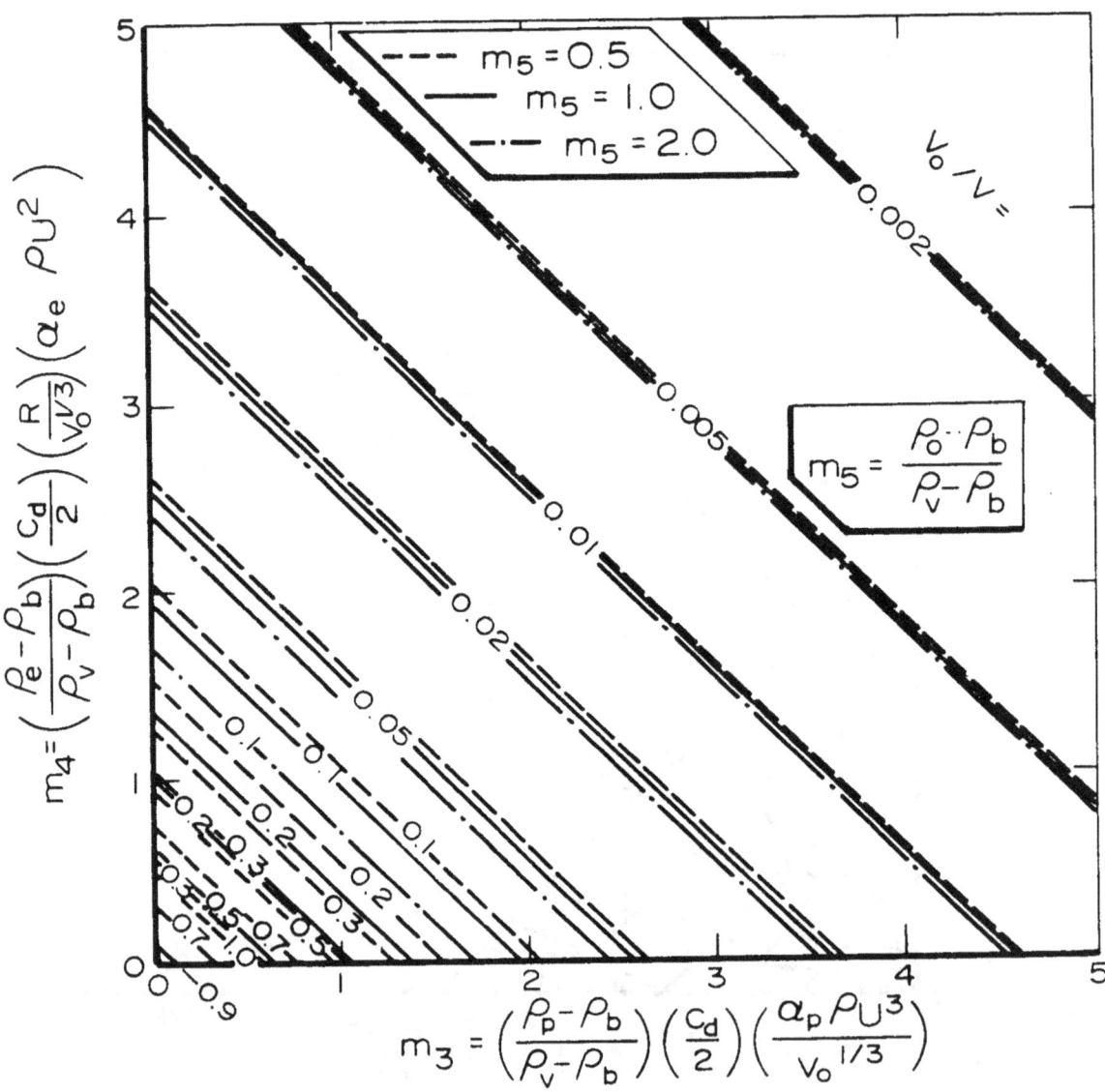

Figure 19 — Vehicle limited by power, energy, and density

Once the preliminary work of evaluating the mission parameters has been completed, the design form can be obtained from the results of the generalized design mission in a matter of minutes.

Power-, Energy-, Density-, and Weight- or Volume-Limited Vehicles

If a vehicle is weight or volume limited in addition to being power, energy, and density limited, the illustration of the generalized design mission changes only by the addition of a boundary line which appears in the (m_3, m_4, m_5) space, and which corresponds to the maximum allowable weight W_{max} or volume V_{max}. The weight and volume limitations are expressed by the values of the following new mission parameters:

$$\text{(weight limited)} \qquad m_6 = \frac{W_{max}}{\rho_o V_o g} = \left(\frac{W}{\rho_o V_o}\right)_{max} \cdot \frac{1}{g} \qquad (92)$$

$$\text{(volume limited)} \qquad m_7 = \frac{V_{max}}{V_o} = \left(\frac{V}{V_o}\right)_{max} \qquad (93)$$

Except for the limiting boundary line in the (m_3, m_4, m_5) space, the theory and solution is the same as in the generalized mission solved in the preceding section. Therefore, Figure 19 is still valid for values of m_3 and m_4 up to the limiting boundary line defined by Equation 93 for a volume restriction, and by the following modification of Equation 92 for a weight restriction:

$$\text{(weight limited)} \qquad m_6 = \left(\frac{\rho V}{\rho_o V_o}\right)_{max} = \frac{\rho}{\rho_o}\left(\frac{V}{V_o}\right)_{max} \qquad (94)$$

and rewriting,

$$\text{(weight limited)} \qquad \left(\frac{V}{V_o}\right)_{max} = \left(\frac{\rho_o}{\rho}\right)m_6 = \frac{W_{max}}{\rho V_o g} \qquad (95)$$

Classification of Submerged Vehicles

Three of the four design parameters V_o/V, V_p/V, V_e/V, and V_b/V are required to classify the form of a submerged vehicle when its density is critical; otherwise, two of the three parameters V_o/V, V_p/V, and V_e/V are required to classify it. These parameters are utilized for technical classification because, for any given category of submerged vehicles with a turbulent boundary layer, the external form does not significantly affect the vehicle performance as long as it is reasonably streamlined as discussed earlier. For a given vehicle, the relative speed, range, and buoyancy requirements are indicated by the relative sizes of the power-, energy-, and buoyancy-dependent components of the vehicle.

The design mission of a submerged vehicle is classified by the parameters m_1 to m_5 if density is important; otherwise, only m_1 and m_2 are required for classification. Although no simple linear relationship exists between the various m's and the design parameters, Figures 17 and 19 and the design equations indicate that in most cases m_1 affects V_p, m_2 affects V_e, m_3 affects W_p, m_4 affects W_e, and m_5 affects V_b and W_b.

A completely different kind of classification parameter, which has no bearing on the design form of submerged vehicles, but permits their performance to be compared with surface craft and airplanes, is the weight-to-drag ratio, W/D. This ratio is relatively invariant for certain kinds of surface craft and air-craft, and determines the amount of thrust required to propel a given vehicle. The value of W/D is around 4 to 6 for well-designed planing boats, 6 to 10 for hydrofoil boats, 10 to 20 for airplanes, 20 to 40 for sailplanes, and up to 100 to 200 for large ships. The value of W/D for submerged vehicles can vary over this entire range, and beyond, depending upon their size and speed. Wislicenus (19) introduces D/W as a function of Froude number, presents a discussion on the use of D/W for classification, and includes a survey of the state of the art of submerged vehicles.

If the value of W/D for a submerged vehicle is low, and the design problem is sufficiently flexible, the designer may want to consider redesigning the vehicle so that it can travel part of the time on the surface or in the air. On the other hand, the designer may find that continuous travel on the surface or in the air is best performancewise, if the design mission is general enough to permit this. An example of a vehicle which travels through both the air and water during operation is the Navy ASROC missile. This missile operates like a rocket initially, and then like a torpedo after shedding one stage and entering the water. The Polaris missile is an alternate example where the first stage of travel occurs underwater, and the next stage takes place in the air. More common examples of vehicles designed for two-phase operation are submarines which are designed also for surface travel.

The expression for the W/D ratio of submerged vehicles is

$$\frac{W}{D} = \frac{\rho g V}{C_d V^{2/3} \frac{1}{2}\rho U^2} = \frac{2}{C_d} \cdot \frac{g V^{1/3}}{U^2} \qquad (96)$$

Notice that W/D is inversely proportional to the square of the Froude number based on volume. Rewriting Equation 96,

$$\frac{W}{D} = \frac{2}{C_d} \cdot \frac{g V_o^{1/3}}{U^2} \left(\frac{V}{V_o}\right)^{1/3} \qquad (97)$$

The parameter $g V_o^{1/3}/U^2$ is seen to be of prime importance for comparing the performance of submerged vehicles with other vehicles. The value of W/D increases as the size V increases and as U reduces. Consequently, the large, low-speed vehicles, like some kinds of research submarines, will have a high value of W/D, while the small high-speed torpedoes will have a low value of W/D. In fact, the W/D ratio of some torpedoes is on the order of one, while W/D for some research submarines is on the order of 100.

Numerical Examples

Example (a). This first design example is the design of a two-man research submarine. The mission specifications are: U = 10 ft/sec; R = 50 miles; z = 10,000 ft; the fluid is sea water where ρ = 2 slugs/ ft^3, ν = 1.2 x 10^{-5} ft^2/sec, and P_v = 30 lbs/ft^2; P = 2120 lbs/ft^2; V_o = 125 ft^3 and ρ_o = 1.5 slugs/ft^3 for a glass sphere, personnel, equipment, flooded compartments, controls, stabilizing surfaces, etc; ρ_b = 1.0 slugs/ft^3, ρ_v = 2.0 slugs/ft^3;

α_p = 2.5 x 10^{-5} ft^3/(ft lb/sec) and ρ_p = 10 slugs/ft^3 for an electric motor, propulsor, and the associated structure; and α_e = 1.4 x 10^{-7} ft^3/ft lb and ρ_e = 10 slugs/ft^3 for the batteries.

The cavitation number is $\sigma = (P-P_v)/\frac{1}{2}\rho U^2$ = 20.9, so there is no cavitation problem. The Reynolds number $U\ell/\nu$ is around 10^7 since ℓ is around 10 feet, so the boundary layer is definitely turbulent. Assuming that a number of protuberances are required for research investigations, that various lifting hooks, railings, bumpers, and structural supports exist, and that the sphere is not completely faired, the drag coefficient would be around C_d = 0.10.

Using Equations 79 to 83, m_1 = 0.00050, m_2 = 0.073, m_3 = 0.0045, m_4 = 0.66, and m_5 = 0.50. Using Figure 19, $V_o/V \doteq 0.8$; the more accurate value of 0.78 is obtained from Equation 88. Using Equations 89 to 91, V_p/V = 0.00046, V_e/V = 0.0673, and V_b/V = 0.1522. The total vehicle volume is 125/0.78 = 160 ft^3, and its submerged weight (with water compartments filled) is W = (160)(2)(32.2) = 10,300 lbs. The volume and weight of the power-dependent components are V_p = (0.00046)(160) = 0.0736 ft^3 and W_p = (10)(32.2)(0.0736) = 24 lbs. The volume and weight of the energy-dependent components are V_e = (0.0673)(160) = 10.8 ft^3 and W_e = 3,480 lbs. The volume and weight of the buoyancy chamber are V_b = 24.4 ft^3, and W_b = (1)(32.2)(24.4)= 785 lbs. The weight of the glass sphere, personnel, instruments, flooded compartments, etc. is W_o = (1.5)(32.2)(125) = 6,040 lbs. The vehicle drag is D = $C_d V^{2/3}\frac{1}{2}\rho U^2$ = 294 lbs., and the power delivered to the water is (294)(10/550) = 5.36 horsepower.

Example (b). The objective is to design a torpedo where the specifications are: $U = 80$ ft/sec, $R = 30,000$ ft, $V_o = 1.0$ ft^3, $\rho_o = 4.4$ slugs/ft^3, $\rho_v = 2.4$ slugs/ft^3, $\nu = 1.2 \times 10^{-5}$ ft lbs/sec, $\rho = 2.0$ slugs/ft^3, $\rho_b = 0.4$ slugs/ft^3, $\alpha_p = 1.5 \times 10^{-5}$ ft^3/(ft lb/sec) for a thermal power plant at operating depth, $\rho_p = 6.2$ slugs/ft^3, $\alpha_e = 1.0 \times 10^{-7}$ ft^3/ft lb, and $\rho_e = 4.2$ slugs/ft^3. The drag coefficient of a smooth, streamlined body with a turbulent boundary layer and stabilizing fins with $R_e = U\ell/\nu \doteq 4 \times 10^7$ using (17) is $C_d = 0.023$.

Using Equations 79 to 83, $m_1 = 0.177$, $m_2 = 0.442$, $m_3 = 0.513$, $m_4 = 0.840$, and $m_5 = 2.0$. Using Figure 19, $V_o/V \doteq 0.14$; a more accurate value is $V_o/V = 0.145$, using Equation 88. Consequently, $V = 6.90$ ft^3 and $W = 534$ lbs. Using Equations 89 to 91, $V_p/V = 0.0929$, $V_e/V = 0.232$, and $V_b/V = 0.470$. Consequently, $V_p = 0.641$ ft^3, $W_p = 128$ lbs, $V_e = 1.60$ ft^3, $W_e = 217$ lbs, $V_b = 3.24$ ft^3, and $W_b = 42$ lbs. Also, $W_o = 142$ lbs. The vehicle drag is $(0.023)(6.90)^{2/3}(80)^2 = 533$ lbs, and the power delivered to the water is $(533)(80)/550 = 77.7$ horsepower.

Example (c). This last example is an airship design where $W_o = g\rho_o V_o = 30,000$ lbs, $U = 150$ ft/sec, $R = 3,000$ mi, $\rho_o = 0.4$ slugs/ft^3, $\rho_b = 0.00119$ slugs/ft^3 (which includes structure which is designed for the given speed), $\alpha_p = 1.5 \times 10^{-5}$ ft^3/(ft lb/sec), $\rho_p = 6$ slugs/ft^3, $\alpha_e = 4 \times 10^{-9}$ ft^3/ft lb, $\rho_e = 2$ slugs/ft^3, $\rho = \rho_v = 0.00238$ slugs/ft^3, and $\nu = 1.4 \times 10^{-4}$.

The Reynolds number is $R_e = U\nu/\ell \doteq 2 \times 10^8$. Using (17), and assigning some drag interference between the passenger compartment and the buoyancy compartment, $C_d \doteq 0.023$.

Using Equations 79 to 83, $m_1 = 0.000105$, $m_2 = 0.00295$, $m_3 = 0.53$, $m_4 = 4.95$, and $m_5 = 335$. Using Equation 88, $V_o/V = 0.00121$. Therefore, $V = 1,920,000$ ft^3 and $W = 147,000$ lbs. Using Equations 89 and 90, $V_p/V = 0.0000111$, $V_e/V = 0.000312$, so $V_p = 21.3$ ft^3, $V_e = 600$ ft^3, $W_p = 4,110$ lbs, and $W_e = 38,600$ lbs. Using Equation 91, $V_b = 1,919,000$ ft^3, so $W_b = 73,500$ lbs. The drag is $D = (0.023)(15,500)(0.00119)(22,500) = 9,530$ lbs, and the net power output is $(9,530)(150)/550 = 2,600$ HP. The ratio $W_o/D = 3.15$, which is somewhat lower than the equivalent value for passenger aircraft. Because of this lower ratio, the lower speed, much greater hanger difficulties, and safety problems, it can be seen why the airship is not competitive with modern commercial jet airplanes.

Effect of Technological Improvements on the Performance and Form of Submerged Vehicles

The solution of the generalized design mission for submerged vehicles permits many questions to be answered regarding the effects of technological improvements on vehicle performance or design form. Such answers aid in deciding whether new drag reduction methods should be applied to torpedoes or submarines, whether research dollars should be spent on miniaturizing the electronics or on improving the energy output of the fuel or batteries, whether the space saved by reducing the payload size should be utilized for increasing the power and speed or for increasing the stored energy and range, how the design form or weight changes if one or more vehicle components is reduced in size, etc.

Change in design form. The quickest method for determining
the effect of a technological improvement on design form is to
first calculate the new values of mission parameters m_1 and m_2 in
case density is unimportant, or m_1 to m_5 in case density is important. Figure 17, or Figure 19 and Equations 89 to 91 are then used
to provide the new design form for each respective case. If greater
accuracy is desired, Equations 85 and 88 can be substituted for
Figures 17 and 19, respectively. By doing this, the effect on
design form can be quickly obtained for any combination of specification changes. The total volume or weight change can be calculated
using both the given data and the design form parameters which result
from the new set of mission parameters.

Speed increase. Another way of utilizing a technological
improvement is to increase the speed of a given vehicle. The
assumptions made for the following analysis are: (a) the vehicle
size, V_o, and V_b remain fixed; and (b) any change in buoyancy is
either small or unimportant. Other assumptions could have been
made, but these are a simple and practical set, and serve to provide
an example of the many kinds of analysis that can be conducted.
Technological changes in α_p, α_e, R, and C_d will be considered.

Consider first, an improvement in the power output, which
enters this problem as a reduction in α_p by the factor α_p/α_{px},
where α_{px} is the improved value. Since V_o/V and V_b/V_o are invariant, Equation 68 reduces to

$$\frac{C_d \rho U^2}{2V_o^{1/3}} (\alpha_p U + \alpha_e R) = m_1 + m_2 = \text{constant} = m_{1x} + m_{2x} \qquad (98)$$

where m_1 and m_2 are given by Equations 79 and 80, and the subscript x refers to the new value which results from any technological change. Rewriting Equation 98 for the change in U which results from changing α_p to α_{px} yields

$$m_1 + m_2 = m_1 \frac{\alpha_{px}}{\alpha_p} \left(\frac{U_x}{U}\right)^3 + m_2 \left(\frac{U_x}{U}\right)^2 \tag{99}$$

Rewriting Equation 99,

$$\frac{\alpha_p}{\alpha_{px}} = \frac{m_1 \left(\frac{U_x}{U}\right)^3}{m_1 + m_2 \left[1 - \left(\frac{U_x}{U}\right)^2\right]} = \frac{\frac{m_1}{m_2}\left(\frac{U_x}{U}\right)^3}{\frac{m_1}{m_2} + 1 - \left(\frac{U_x}{U}\right)^2} \tag{100}$$

From Equations 86 and 87, it is seen that

$$\frac{m_1}{m_2} = \frac{V_p}{V_e} \tag{101}$$

Substituting Equation 101 into Equation 100 provides

$$\frac{\alpha_p}{\alpha_{px}} = \frac{\frac{V_p}{V_e}\left(\frac{U_x}{U}\right)^3}{\frac{V_p}{V_e} + 1 - \left(\frac{U_x}{U}\right)^2} \tag{102}$$

which shows the improvement factor in α_p that is required to increase vehicle speed by a factor of U_x/U. Since the result is a function only of V_p/V_e, then any type, form, or size of submerged vehicle having the same ratio of V_p/V_e will benefit equally (from the viewpoint of speed increase factor) from a given improvement in α_p. Consequently, V_p/V_e (or m_1/m_2) may be viewed as a special type of classification parameter.

Utilizing Equation 98, and following the same procedure, the factors α_e/α_{ex}, R/R_x, and C_d/C_{dx}, required for a given speed increase, are found to be

$$\frac{\alpha_e}{\alpha_{ex}} = \frac{\left(\dfrac{U_x}{U}\right)^2}{1 + \dfrac{V_p}{V_e}\left[1 - \left(\dfrac{U_x}{U}\right)^3\right]} \qquad (103)$$

$$\frac{R}{R_x} = \frac{\left(\dfrac{U_x}{U}\right)^2}{1 + \dfrac{V_p}{V_e}\left[1 - \left(\dfrac{U_x}{U}\right)^3\right]} \qquad (104)$$

$$\left(\frac{C_d}{C_{dx}}\right)_o = \frac{\left(\dfrac{U_x}{U}\right)^2\left[\dfrac{V_p}{V_e}\dfrac{U_x}{U} + 1\right]}{1 + \dfrac{V_p}{V_e}} \qquad (105)$$

where the parenthesis around C_d/C_{dx} with the sub-o means that it is assumed that the drag reduction method requires no additional equipment.

Equations 102 to 105 are plotted in Figure 20 showing the improvement factor which is needed for a speed increase of 25%. Similar curves could be plotted for other speed increases. Notice that drag reduction appears best, since lower improvement factors are needed. Also, notice that a considerable improvement in α_e or

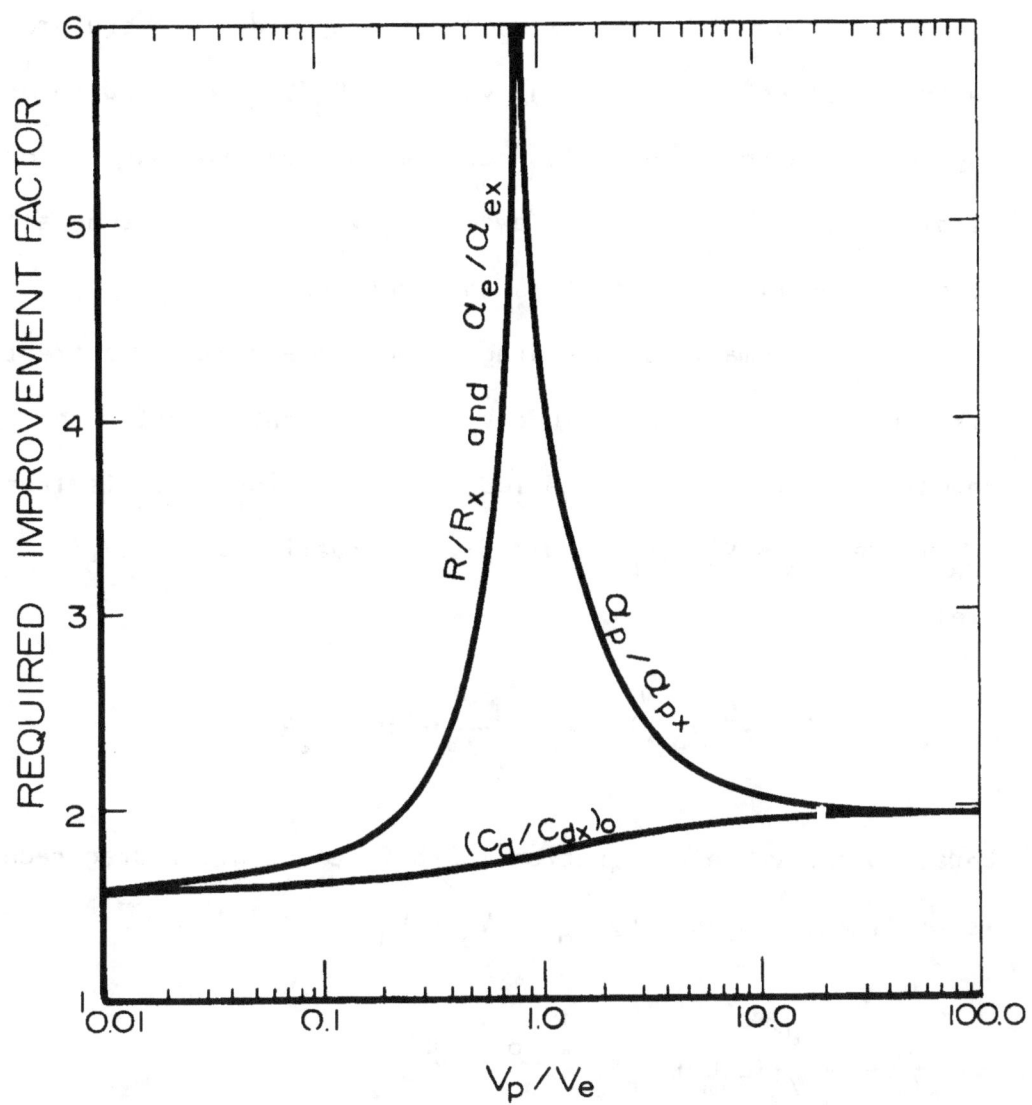

Figure 20 - Improvement factor required for a speed increase of 25%

range is required unless the ratio V_p/V_e is less than about 0.6. Similarly, the gain resulting from an improvement in α_p is small unless V_p/V_e is greater than about 1.5.

If internal volume is used for drag reduction equipment (which is often the case), the value of C_d/C_{dx} required to provide a given speed increase would increase in order to make up for the usable volume which is taken away from V_p and V_e. The penalty due to drag reduction equipment is analyzed next.

The volume V_d of the drag reduction equipment is treated like an increase in the volume V_o. Consequently, after the drag reduction equipment is installed, the new value of V_o is labeled V_{ox} where $V_{ox} = V_o + V_d$. Multiplying Equation 68 by $(V_o/V)^{1/3}$ yields

$$1 - \frac{V_b}{V} = \frac{V_o}{V} + \frac{C_d \rho U^2}{2V^{1/3}}(\alpha_p U + \alpha_e R) = \text{constant} \qquad (106)$$

Equating the value of Equation 106 before and after drag reduction is utilized, and letting $V_{ox} = V_o + V_d$, yields

$$\frac{V_o}{V} + \frac{C_d \rho U^2}{2V^{1/3}}(\alpha_p U + \alpha_e R) = \frac{V_o + V_d}{V} +$$

$$[\alpha_p U \left(\frac{U_x}{U}\right) + \alpha_e R] \frac{C_d \rho U^2}{2V^{1/3}} \left(\frac{U_x}{U}\right)^2 \frac{C_{dx}}{C_d} \qquad (107)$$

Solving for C_d/C_{dx}, Equation 107 becomes

$$\frac{C_d}{C_{dx}} = \frac{\dfrac{C_d \rho U^2}{2V^{1/3}} \left(\dfrac{U_x}{U}\right)^2 \alpha_p U \left(\dfrac{U_x}{U}\right) + \alpha_e R}{\dfrac{C_d \rho U^2}{2V^{1/3}} (\alpha_p U + \alpha_e R) - \dfrac{V_d}{V}} \tag{108}$$

Using Equations 79 and 86, it is seen that

$$\frac{\alpha_p C_d \rho U^3}{2V^{1/3}} = m_1 \left(\frac{V_o}{V_1}\right)^{1/3} = \frac{V_p}{V} \tag{109}$$

$$\frac{\alpha_e C_d \rho U^2 R}{2V^{1/3}} = m_2 \left(\frac{V_o}{V}\right)^{1/3} = \frac{V_e}{V} \tag{110}$$

Substituting Equations 109 and 110, Equation 108 becomes

$$\frac{C_d}{C_{dx}} = \frac{\dfrac{V_p}{V}\left(\dfrac{U_x}{U}\right)^3 + \dfrac{V_e}{V}\left(\dfrac{U_x}{U}\right)^2}{\dfrac{V_p}{V} + \dfrac{V_e}{V} - \dfrac{V_d}{V}} \tag{111}$$

If $V_d = 0$, then by definition, Equation 111 becomes equal to $(C_d/C_{dx})_o$, or

$$\left(\frac{C_d}{C_{dx}}\right)_o = \frac{\dfrac{V_p}{V}\left(\dfrac{U_x}{U}\right)^3 + \dfrac{V_e}{V}\left(\dfrac{U_x}{U}\right)^2}{\dfrac{V_p}{V} + \dfrac{V_e}{V}} \tag{112}$$

which is the same as Equation 105. Dividing Equation 111 by Equation 112 yields

$$\frac{C_d}{C_{dx}} = \left(\frac{C_d}{C_{dx}}\right)_o \frac{\frac{V_p}{V} + \frac{V_e}{V}}{\frac{V_p}{V} + \frac{V_e}{V} - \frac{V_d}{V}} \qquad (113)$$

Substituting Equation 91,

$$\frac{C_d}{C_{dx}} = \left(\frac{C_d}{C_{dx}}\right)_o \frac{1}{1 - \frac{V_d}{V - V_o - V_b}} \qquad (114)$$

Equation 114 shows that the actual drag reduction factor C_d/C_{dx}, when drag reduction equipment is needed, must be greater than the drag reduction factor $(C_d/C_{dx})_o$ which is required when no added equipment is assumed. This penalty is seen from Equation 114 to increase as V_d/V increases, and as V_o/V and V_b/V increase.

At this point, it is interesting to remove the criterion that V_o must remain fixed, and determine how much the speed of a vehicle might be increased if the size of the payload, electronics components, and the associated structure is reduced. Equating the value of Equation 106 before and after a change in V_o is made, yields

$$\frac{V_o}{V} + \frac{C_d \rho U^2}{2V^{1/3}}(\alpha_p U + \alpha_e R) = \left(\frac{V_o}{V}\right)\left(\frac{V_{ox}}{V_o}\right) + \frac{C_d \rho U^2}{2V^{1/3}}\left(\frac{U_x}{U}\right)^2 \left[\alpha_p U \left(\frac{U_x}{U}\right) + \alpha_e R\right]$$

$$\tag{115}$$

Rewriting,

$$\frac{V_o}{V_{ox}} = \frac{\dfrac{V_o}{V}}{\dfrac{V_o}{V} - \dfrac{\alpha_p C_d \rho U^3}{2V^{1/3}}\left[\left(\dfrac{U_x}{U}\right)^3 - 1\right] - \dfrac{\alpha_e C_d \rho U^2 R}{2V^{1/3}}\left[\left(\dfrac{U_x}{U}\right)^2 - 1\right]} \tag{116}$$

After substituting Equations 109 and 110, Equation 116 becomes

$$\frac{V_o}{V_{ox}} = \frac{\dfrac{V_o}{V}}{\dfrac{V_o}{V} - \dfrac{V_p}{V}\left[\left(\dfrac{U_x}{U}\right)^3 - 1\right] - \dfrac{V_e}{V}\left[\left(\dfrac{U_x}{U}\right)^2 - 1\right]} \tag{117}$$

This equation is not graphed along with the others in Figure 20 since it is not a sole function of V_p/V_e. The required reduction in V_o for a given speed gain is seen to increase as V_p/V and V_e/V increase, and as V_o/V reduces.

Range increase. In this section, it is assumed that a technological improvement is used to increase the range of a vehicle. The following assumptions are made in the analysis: (a) the vehicle size, V_o, and V_b are fixed; and (b) any change in buoyancy of the vehicle is either small or unimportant. Technological changes in α_p, α_e, U, and C_d are considered.

The application of Equation 98 to the case of range increase resulting from a reduction (improvement) in α_p yields

$$\frac{C_d \rho U^2}{2V_o^{1/3}} (\alpha_p U + \alpha_e R) = \frac{C_d \rho U^2}{2V_o^{1/3}} [\alpha_p U \frac{\alpha_{px}}{\alpha_p} + \alpha_e R (\frac{R_x}{R})] \qquad (118)$$

Solving for R_x/R,

$$\frac{R_x}{R} = \frac{\alpha_p U}{\alpha_e R} (1 - \frac{\alpha_{px}}{\alpha_p}) + 1 \qquad (119)$$

Dividing Equation 73 by Equation 74 yields

$$\frac{\alpha_p U}{\alpha_e R} = \frac{V_p}{V_e} \qquad (120)$$

Substituting Equation 120 into Equation 119 gives the range increase factor as a function of α_p/α_{px} and V_p/V_e where

$$\frac{R_x}{R} = 1 + \frac{V_p}{V_e} (1 - \frac{\alpha_{px}}{\alpha_p}) \qquad (121)$$

Notice that the maximum possible range increase factor is $1 + V_p/V_e$. Proceeding in a similar manner for the other kinds of technological improvement, it can be shown that

$$\frac{R_x}{R} = \frac{U}{U_x}^2 \left\{ 1 + \frac{V_p}{V_e} [1 - (\frac{U_x}{U})^3] \right\} \qquad (122)$$

$$\frac{R_x}{R} = \frac{\alpha_e}{\alpha_{ex}} \tag{123}$$

and

$$\frac{R_x}{R} = \left(\frac{C_d}{C_{dx}}\right)_o \left(1 + \frac{V_p}{V_e}\right) - \frac{V_p}{V_e} \tag{124}$$

where $\left(C_d/C_{dx}\right)_o$ is defined as the drag reduction factor when no

drag reduction equipment is needed.

Equations 121 to 124 are graphed in Figure 21 which presents

the range increase factor R_x/R as a function of the improvement

factors for α_p, α_e, U, and C_d for the cases of $V_p/V_e = \frac{1}{2}$ and $V_p/V_e =$

2. Figure 21 has been graphed differently from Figure 20 to

illustrate an alternate type of presentation. Notice that an

improvement in α_p by a factor of more than two does not help

appreciably. Drag reduction is seen to be even more beneficial than

α_e/α_{ex} because the range factor increases faster than the drag

reduction factor. The greatest range increase results from reducing

speed, but this is seldom feasible unless technological improvements

in the overall system which utilizes the submerged vehicle permit the

vehicle speed to be reduced.

In order to calculate the penalty due to the added volume

V_d of drag reduction equipment, the constant in Equation 106 is

equated before and after the application of drag reduction, where

$V_{ox} = V_o + V_d$, giving

141

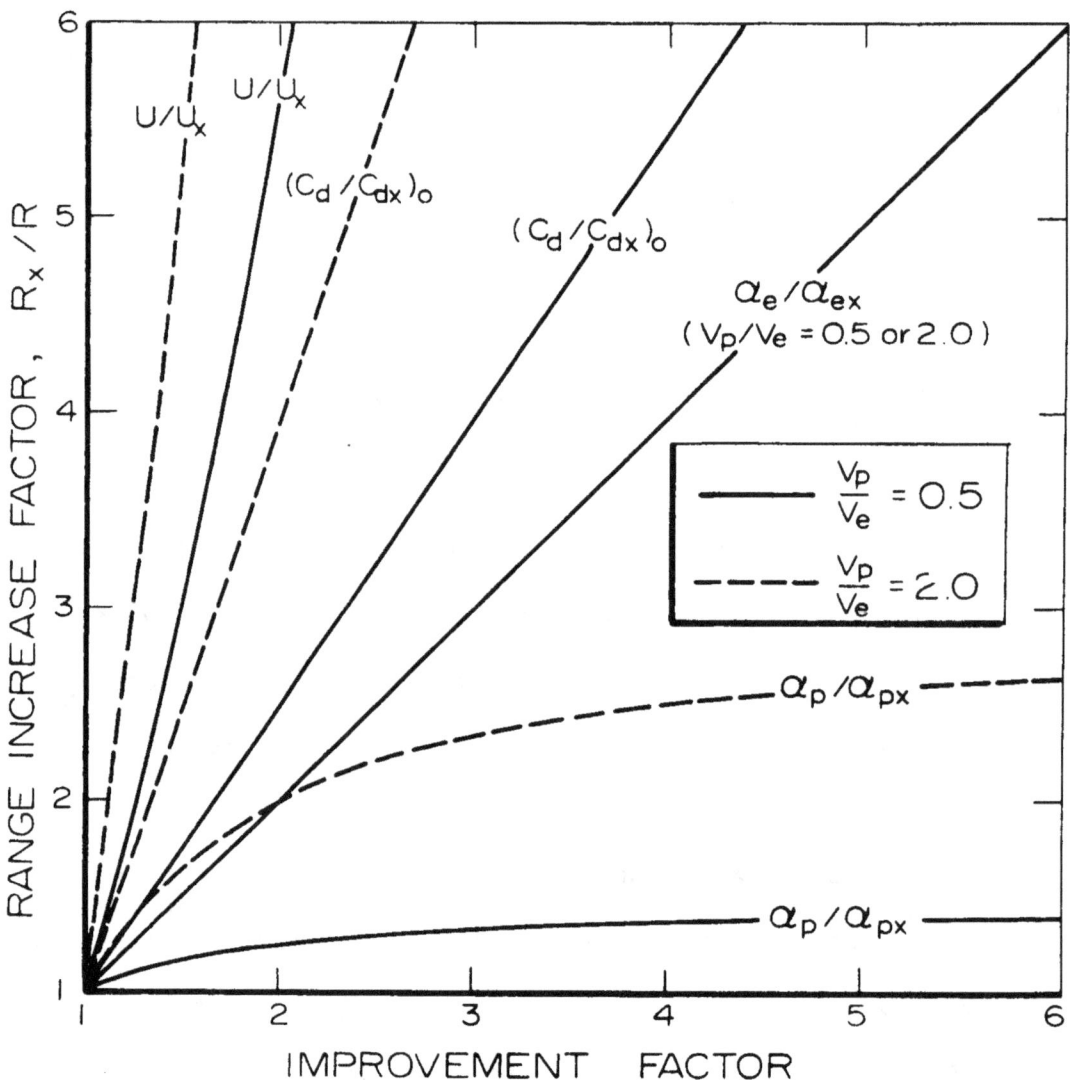

Figure 21 - Effect of technological improvements on
increasing the range

$$\frac{V_o}{V} + \frac{C_d \rho U^2}{2V^{1/3}}(\alpha_p U + \alpha_e R) = \frac{V_o + V_d}{V} + \frac{C_d \rho U^2}{2V^{1/3}}\left(\frac{C_{dx}}{C_d}\right)\left[\alpha_p U + \alpha_e R\left(\frac{R_x}{R}\right)\right] \tag{125}$$

Solving for C_d/C_{dx},

$$\frac{C_d}{C_{dx}} = \frac{\dfrac{C_d \rho U^2}{2V^{1/3}}\left[\alpha_p U + \alpha_e R\left(\frac{R_x}{R}\right)\right]}{\dfrac{C_d \rho U^2}{2V^{1/3}}(\alpha_p U + \alpha_e R) - \dfrac{V_d}{V}} \tag{126}$$

Using the same general method as in developing Equation 114 from Equation 111, Equation 126 yields

$$\frac{C_d}{C_{dx}} = \left(\frac{C_d}{C_{dx}}\right)_o \frac{1}{1 - \dfrac{V_d}{V - V_o - V_b}} \tag{127}$$

Notice that Equation 127 is the same as Equation 114; thus, the penalty for drag reduction equipment is the same whether speed or range is to be increased.

The requirement that V_o is constant will now be removed in order to determine how a reduction in V_o affects range. Equating the constant in Equation 106 before and after a reduction in V_o to increase range yields

$$\frac{V_o}{V} + \frac{C_d \rho U^2}{2V^{1/3}} (\alpha_p U + \alpha_e R) = \left(\frac{V_o}{V}\right)\left(\frac{V_{ox}}{V_o}\right) + \frac{C_d \rho U^2}{2V^{1/3}} \left[\alpha_p U + \alpha_e R \left(\frac{R_x}{R}\right)\right] \qquad (128)$$

Solving for R_x/R, and substituting Equation 110 gives

$$\frac{R_x}{R} = 1 + \frac{V_o}{V_e} \left(1 - \frac{V_{ox}}{V_o}\right) \qquad (129)$$

Notice that the range factor increases as V_{ox}/V_o and V_e/V reduce, and as V_o/V increases.

This concludes the work on the effect of technological changes on the form and performance of submerged vehicles. A different approach to the design of torpedoes was developed by Brumfield (20) which may be of interest to the reader since it shows the effect of design parameters on torpedo size and weight, includes methods for minimizing torpedo weight and volume as a function of target range, and permits the calculation of an optimum speed ratio with respect to the target.

CHAPTER VI

SUMMARY AND CONCLUSIONS

Summary

The objective of this study was to develop a generalized engineering design procedure based on a nondimensional analytical approach which can be used to solve a set of design problems in a given field. An optimization criterion is incorporated into the design procedure. The use of the procedure aids in determining the diversity of design form families, the variations of design form within a family, the relationship between design form and the design objective, the scaling and classification of design forms, the areas where invention is still needed, and the nature of research studies which might lead to new information on design forms.

The generalized design procedure consists essentially of seven steps which are described in Chapter III and consist of: (1) generalize a typical design problem, (2) determine possible design forms, (3) introduce physical relationships, (4) select the mission and design parameters, (5) specify the mapping criteria and the design equations, (6) select a sequence of subspaces to map from mission space, and (7) map from mission space to design space.

Many aspects of the design procedure are discussed in detail. Some of these aspects are: (a) procedure for generalizing a design problem, (b) methods for nondimensionalizing design problems and designs, (c) selection of design mission parameters, (d) establishment of an optimization criterion, (e) treatment of design problems which have a large number of variables, (f) selection of subdesign problems, (g) means for finding typical design forms, (h) relationship between families of design forms and sets of design missions, (i) means for establishing physical relationships between the mission parameters and the design form parameters, (j) uniqueness between the number of mapping relations and the number of design form parameters, (k) methods for illustrating the design form solutions, (l) development of means for classifying design forms and design missions, (m) methods for developing scaling laws, (n) modification of the design procedure for use in research, and (o) selection of design variables for research studies.

Several examples are presented to illustrate the use of the design procedure. Some examples relate to complete design problems while others relate to subdesign problems. The design examples presented are the following: (a) circular tubes under external pressure, (b) columns under compressive loads, (c) submerged vehicles, (d) airplane wings and fully-wetted hydrofoils, and (e) cavitating and noncavitating hydrofoil cross sections. The design procedure is also applied to solve an economic problem in order to illustrate its use in fields other than engineering design.

The results of the design study on submerged vehicles are applicable to the design of such vehicles as naval submarines, small research submarines, torpedoes, remote-controlled underwater vehicles, and airships. The effects of range, speed, type of propulsion system

and fuel, payload size and weight, depth, and buoyancy requirements
on the size and weight of the vehicle and its component parts are
determined. Also, the effect of technological improvements on design
form and performance is determined. Simple means for classifying all
vehicles and design missions are developed. The effect of drag
reduction is compared with the effect of other kinds of improvements.

The design study of airplane wings and hydrofoils results in
the determination of the optimum aspect ratio, thickness ratio,
planform taper, thickness ratio taper, cross-sectional shape, and
lift coefficient. The design problem variables include the weight
of various components, design stress, speed, fluid density and
viscosity, etc. In the case of hydrofoils, the effects of cavitation
number, divergence, and flutter are considered in addition to strength
and viscosity. New scaling and classification parameters are
developed.

The design example of hydrofoil cross sections includes the
effects on design form of such variables as lift, bending moment,
cavitation number, speed, structural strength, fluid density, and
fluid viscosity. Six different families of design forms result,
each of which are associated with a specific set of design missions.
The optimum design form is determined for each of the specific design
missions. Five of the design form families are cavitating hydrofoil
cross sections. A new classification parameter is developed which
permits all cavitating hydrofoils and the simpler fully-wetted
hydrofoils to be classified much like the specific speed parameter

classifies turbomachinery; however, the hydrofoil classification parameter is more general because it also includes the effects of cavitation and structural strength on design form.

Conclusions

The following conclusions are drawn:

1. The generalized engineering design procedure can be applied to a wide variety of engineering design fields. It can also be applied to fields outside of engineering design, and to research studies.

2. The design procedure permits an entire design field to be more completely understood, and shows how design forms are related to design missions.

3. The use of the design procedure may lead to the discovery of new design forms, new families of design forms, and new classification and scaling parameters. Also, it permits the establishment of areas in which inventions or research studies are still needed.

4. The design procedure is a method which aids in solving either simple design problems or complex design problems with many variables. It permits the complex design problems to be more easily treated.

5. Knowledge of the design field and ingenuity are necessary in making use of the full potential of the design procedure.

6. This design procedure is not to be considered the only approach or best approach to design, but merely as a step in the

148

evolution of engineering design theory which hopefully will be of use in its further development. A possible approach to the improvement of design theory may be the development of a more rigorous mathematical foundation.

BIBLIOGRAPHY

1. Mandel, P. "Optimization Methods Applied to Ship Design," The Society of Naval Architects and Marine Engineers, Paper No. 6, Advance Copy for presentation at the Annual Meeting, New York, N.Y., November 10-11, 1966.

2. Zwicky, Fritz. Morphological Astronomy. Berlin: Springer-Verlog, 1957.

3. McLean, W. B. "The Art of Simple and Reliable Design," Office of the Technical Director, U.S. Naval Ordnance Test Station. China Lake, California, Spring, 1963.

4. Gabrielli, G., and Th. von Kármán. "What Price Speed?," Mechanical Engineering, October, 1950, 775-781.

5. Davidson, K. S. M. "Ships," Experimental Towing Tank Technical Memorandum No. 116, Stevens Institute of Technology. Hoboken, New Jersey, September, 1956.

6. Wislicenus, G. F. "Form Design in Engineering," (A Report to the National Science Foundation), The Pennsylvania State University. University Park, Pennsylvania, August 1967.

7. Stahl, W. R. Physiological Similarity and Modelling. New York: Appleton-Century-Crofts, Inc., To be published late in 1968.

8. Werner, R. A. "Analysis of Airplane Design by Similarity Considerations," Master's Thesis, Department of Aerospace Engineering, The Pennsylvania State University. University Park, Pennsylvania, September, 1967.

9. Wislicenus, G. F. Fluid Mechanics of Turbomachines. Volume I and II. New York, N.Y.: Dover Publications, Inc., 1965.

10. Smith, E. Q., Jr., and King, E. F. "Mission Success," U.S. Naval Missile Center, Publication No. NMC-MP-65-12. Point Mugu, California, 10 February, 1966.

11. Timoshenko, S. Strength of Materials. Part II. New York, N.Y: D. Van Nostrand Company, Inc., 1940.

12. Kline, S. J. Similtude and Approximation Theory. New York: McGraw-Hill Book Company, 1965.

13. Sedov, L. I. Similarity and Dimensional Methods in Mechanics. Translated by M. Friedman. New York: Academic Press, 1959.

14. Buckingham, E. "On Physically Similar Systems: Illustrations of the Use of Dimensional Equations," E. Phys. Rev., IV (1914), 345.

15. Munzer, H., and Reichardt, H. "Rotationally Symmetrical Source-Sink Bodies with Predominately Constant Pressure Distribution," Translated by A. H. Armstrong, Armament Research Establishment, Translation No. 1/50. Fort Halstead, Kent, England, April, 1950.

16. Rouse, H., and McNown, J. S. "Cavitation and Pressure Distribution, Head Forms at Zero Angle of Yaw," Bulletin 32, Iowa Institute of Hydraulic Research. Iowa City, Iowa: IIHR, 1948.

17. Brooks, J. D., and Lang, T. G. "Hydrodynamic Drag of Torpedoes," NAVORD Report 5842, U.S. Naval Ordnance Test Station. China Lake, California, 18 February, 1958.

18. Hoerner, S. F. Fluid-Dynamic Drag. Midland Park, N.J., 1965. (Published by author.)

19. Wislicenus, G. F. "Hydrodynamics and Propulsion of Submerged Bodies," American Rocket Society Journal, December, 1960 , 1140-1148.

20. Brumfield, R. C. "Factors Influencing the Size and Weight of Underwater Vehicles," American Rocket Society Journal, December, 1960, 1152-1160.

21. Abbott, I. H., and Von Doenhoff, A. E. Theory of Wing Sections, New York: Dover Publications, Inc., 1959.

22. Abramson, H. N., Chu, W. H., and Irick, J. T. "Hydroelasticity," Monograph draft copy, Southwest Research Institute. San Antonio, Texas, August, 1966.

23. Bisplinghoff, R. L., Ashley, H., and Halfman, R. L. Aeroelasticity. Reading, Massachusetts: Addision-Wesley Publishing Company, Inc., 1955.

24. Timoshenko, S., and MacCullough, G. H. Elements of Strength of Materials. New York: D. Van Nostrand Company, Inc., 1940.

25. McCormick, B. W., Jr. Aerodynamics of V/STOL Flight. New York: Academic Press, 1967.

26. Tulin, M. P., and Burkart, M. P. "Linearized Theory for Flows About Lifting Foils at Zero Cavitation Number," Report C-683, The David Taylor Model Basin. Washington, D. C., February, 1955.

27. Tulin, M. P. "Steady Two-Dimensional Cavity Flows About Slender Bodies," Report 834, The David Taylor Model Basin. Washington, D. C., May, 1953.

28. Lang, T. G. "Base-Vented Hydrofoils," NAVORD Report 6606, U.S. Naval Ordnance Test Station. China Lake, California, 19 October, 1959.

29. Tulin, M. P. "The Shape of Cavities in Supercavitating Flows," Technical Report 121-5 Hydronautics, Inc., Laurel, Md., April, 1965.

30. Johnson, V. E., Jr. "Theoretical Determination of Low-Drag Supercavitating Hydrofoils and Their Two-Dimensional Characteristics at Zero Cavitation Number," NACA RML57G11a, September, 1957.

31. Auslaender, J. "The Linearized Theory for Supercavitating Hydrofoils Operating at High Speeds Near a Free Surface," Journal of Ship Research, October, 1962, 8-23.

32. Auslaender, J. "Low Drag Supercavitating Hydrofoil Sections," Technical Report 001-7, Hydronautics, Inc., Laurel, Md., April, 1962.

33. Tulin, M. P. "Supercavitating Flows - Small Perturbation Theory," Technical Report 121-3, Hydronautics, Inc., Laurel, Md., September, 1963.

34. Wu, T. Y. "A Free Streamline Theory for Two-Dimensional Fully Cavitating Hydrofoils," J. Math. Phys., XXXV (1956), 236-65.

35. Wu, T. Y. "A Note on the Linear and Nonlinear Theories for Fully Cavitated Hydrofoils," California Institute of Technology Hydro. Lab. Report. Pasadena, California, 1956.

36. Wu, T. Y. "A Wake Model for Free-Streamline Flow Theory; Part I, Fully and Partially Developed Wake Flows and Cavity Flows Past an Oblique Flat Plate," J. Fluid Mech, XIII (1962), 161-81.

37. Fabula, A. G. "Application of Thin-Airfoil Theory to Hydrofoils with Cut-Off Ventilated Trailing Edge," NAVWEPS Report 7571, U.S. Naval Ordnance Test Station. China Lake, California, 13 September, 1960.

38. Fabula, A. G. "Linearized Theory of Vented Hydrofoils," NAVWEPS Report 7637, U.S. Naval Ordnance Test Station. China Lake, California, 7 March 1961.

APPENDIX A

DESIGN OF LOW-SPEED AIRPLANE WINGS AND
NONCAVITATING HYDROFOILS

Most complex design problems must be separated into sub-design problems in order to solve them most efficiently. The objective of this appendix is to illustrate how one particular type of design problem is separated into subdesign problems, how one of these subdesign problems is transformed into a generalized design mission, and how the design procedure is used to solve the resulting set of design missions.

The design of dynamically-supported vehicles such as airplanes and hydrofoil boats is selected as the particular type of design problem. The design of a lifting surface is the selected subdesign problem to be solved wherein the lifting surface represents either an airplane wing or a lifting hydrofoil of a hydrofoil boat. The design of lifting surfaces is intended to be as general as possible.

Specification of the Subdesign Problem

In general, the design objective of either an airplane or a hydrofoil boat is to transport a given payload at a given speed for a given distance. All of the vehicle components must be packaged into a dynamically-stable vehicle whose weight is supported by a lifting surface. The lifting surface must be structurally sound and should provide the desired lift with a minimum of drag.

The design problem of an airplane or a hydrofoil boat, like most design problems, is typically an iterative process where the values of certain items are assumed which then permits an analysis to be conducted. The results of the analysis are then used to obtain better values for these certain items so an improved analysis can be conducted.

Utilizing this iterative process, it is assumed that a rough estimate is first made of the weight of all vehicle components. The components include such items as the payload, crew, structure, electronics, power source, fuel, control devices, special equipment, the necessary passenger accommodations, etc. In the case of airplanes, considerable weight can be carried in or on the wing in order to reduce both the bending moment exerted on the wing and its structural weight, and to also reduce the size and drag of the hull. Such items as the fuel, power plant, landing gear, and certain kinds of payload can be placed in or on the wing. Let the estimated value of the weight of all of the components, except the structural weight of the lifting surface, be designated as W_a. Also, let the portion of W_a which is to be placed in or on the lifting surface be called W_x.

The design of the lifting surface can now be considered as a separate design problem. Letting W_s be the structural weight of the lifting surface, the required lift is $W_a + W_s$. The load contributing to the bending moment on the lifting surface is $W_a - W_x$. The total vehicle drag D is equal to $D_a + D_\ell$, where D_ℓ is the drag of the lifting surface, and D_a is the drag of all other components. Let the optimization criterion for the vehicle design problem be to minimize D. Since D_a is essentially fixed and independent of the lifting surface design, the equivalent optimization criterion for the lifting surface subdesign problem is to minimize D_ℓ.

The general design objective of the lifting surface subdesign problem is to determine all of the fundamental variables, which include: planform shape, aspect ratio, chordwise thickness distribution, camber, thickness-to-chord ratio and its spanwise taper, void area of the cross-section, structural weight, drag, and lift-to-drag ratio. Much of the design analysis pertains to both wings and hydrofoils. Since the entire problem is too lengthy to complete here, a point in the analysis will be reached where only hydrofoils are considered.

In case an airplane or a hydrofoil boat has more than one lifting surface, W_a is defined as the portion of the total load carried by a particular lifting surface minus its structural weight.

Generalized Design Mission of the Lifting Surface Problem

The lifting surface problem is now generalized into a set of design missions.

Design problem specifications and mission parameters. The selected set of design problem specifications is: W_a, W_x, speed U, acceleration of gravity g, fluid characteristics which consist of density ρ, kinematic viscosity ν, speed of sound a, pressure P,

vapor pressure P_v, and characteristics of the structural material
which are density ρ_s, modulus of elasticity E, and the design bending
stress f (which includes the load factor and factor of safety).
Summarizing, the set of twelve selected design problem specifications
consists of W_a, W_x, U, g, ρ, ν, a, P, P_v, ρ_s, E, and f. The pi theorem
predicts nine nondimensional parameters. One possible set of
(nondimensional) mission parameters is W_x/W_a, $W_a g^2/\rho U^6$, $U^3/g\nu$, U/a,
$P/\frac{1}{2}\rho U^2$, $P_v/\frac{1}{2}\rho U^2$, ρ_s/ρ, f/E, and $\frac{1}{2}\rho U^2/f$.

Mission criteria. The general design objective is to
determine the form of the lifting surface which has lowest drag.
The nondimensional optimization criterion Q is

$$Q = \frac{D_\ell}{W_a} \qquad (130)$$

where Q is to be minimized.

The following mission criteria are selected for the generalized
lifting surface design mission: (a) steady state operating conditions,
(b) bending stress is the only critical stress problem, (c) all
hydrofoils are noncavitating, (d) the cross-sectional shape of the
lifting surface is constant along the span, and (e) the boundary layer
is turbulent.

Possible Design Forms

Some possible design forms are shown in Figures 22 and 23.
After viewing the possible forms, it is realized that selection of a
lifting surface form is dependent not only on the technical

156

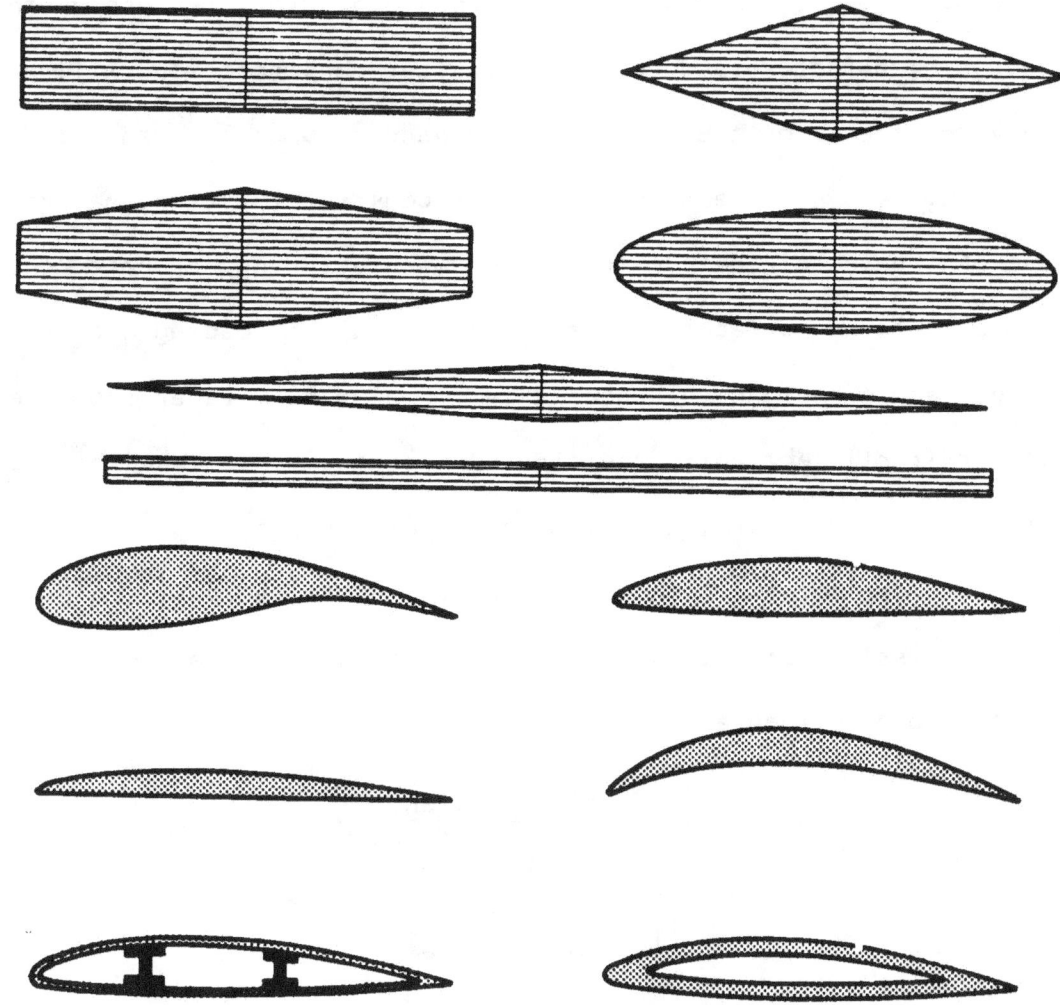

Figure 22 — Possible planforms, thickness tapers, and cross-sectional shapes of lifting surfaces

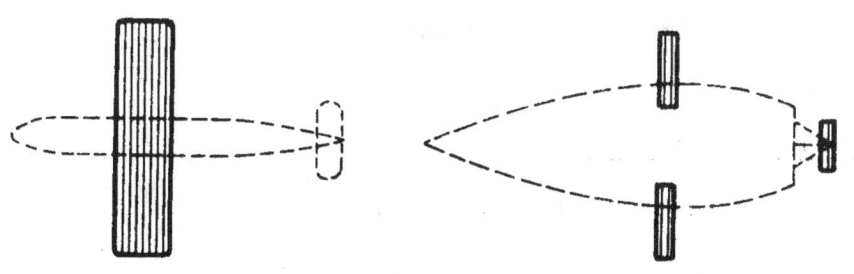

Figure 23 — Examples of lifting surface applications

157

consideration of low drag but also upon other factors such as the ease of manufacture and cost. The latter considerations will have varying effects on the design form depending upon the manufacturing facilities, available development and construction time, nature and use of the vehicle, etc. In view of manufacturing and cost considerations, lifting surfaces generally have straight leading and trailing edges, and either no thickness taper or a uniform taper from the center (root) section to the tip. Consequently, to make this design mission more definite, the planform taper ratio τ, and the thickness-to-chord taper ratio ϕ, are here considered to be specified parameters (i.e., mission parameters) rather than design form parameters, where τ and ϕ are defined as

$$\tau = \frac{\text{tip chordlength}}{\text{root chordlength}}$$

and

$$\phi = \frac{\text{tip t/c ratio}}{\text{root t/c ratio}}$$

Physical Relationships

Optimization criterion Q. The physical equations needed to evaluate Q are related to gravitational and aerodynamic forces. Recalling that the lift L is equal to $W_a + W_s$, Equation 130 is rewritten as

$$Q = \frac{D_\ell}{W_a} = \frac{D_\ell}{L} \cdot \frac{L}{W_a} = \frac{C_d}{C_L} \left(1 + \frac{W_s}{W_a}\right) \tag{131}$$

where the drag coefficient C_d and the lift coefficient C_L are defined as follows:

$$D_\ell = C_d \cdot bc \cdot \tfrac{1}{2}\rho U^2 \tag{132}$$

$$L = C_L \cdot bc \cdot \tfrac{1}{2}\rho U^2 \tag{133}$$

where b is the span of the lifting surface and c is the mean chord-length defined as the planform area divided by b.

The term W_s/W_a in Equation 131 is evaluated by first introducing the relationship

$$W_s = \rho_s g C_v bct \tag{134}$$

where ρ_s is the density of the structural material, $t = (t/c)c$ where (t/c) is defined as the thickness-to-chord ratio at the center of the semispan, and C_v is defined as the volume of structural material divided by bct. Substituting bc from Equation 133 into Equation 134 and rearranging,

$$\frac{W_s}{L} = 2\,\frac{C_v}{C_L}\,\frac{\rho_s}{\rho}\,\frac{gt}{U^2} = \frac{W_s}{W_a + W_s} = \frac{W_s}{W_a}\left(\frac{1}{1 + \dfrac{W_s}{W_a}}\right)$$

Let the aspect ratio A_r be defined as

$$A_r = \frac{b}{c} \tag{135}$$

Setting $b = A_r c$, and $c = t/(t/c)$, Equation 134 can be solved for t and substituted into the expression for W_s/L to give

$$\frac{W_s}{W_a}\left(\frac{1}{1+\frac{W_s}{W_a}}\right) = 2\,\frac{C_v}{C_L}\frac{\rho_s}{\rho}\left(\frac{W_s\left(\frac{t}{c}\right)^2}{\rho_s g C_v A_r}\right)^{1/3}\frac{g}{U^2}\left(\frac{W_a}{W_a}\right)^{1/3}$$

$$= \frac{2C_v}{C_L A_r^{1/3}}\left(\frac{\rho_s}{\rho}\right)^{2/3}\left(\frac{W_s}{W_a}\right)^{1/3}\left(\frac{t}{c}\right)^{2/3}\left(\frac{W_a g^2}{\rho U^6}\right)^{1/3}$$

Rearranging, and raising the equation to the three-halves power,

$$\frac{\frac{W_s}{W_a}}{\left(1+\frac{W_s}{W_a}\right)^{3/2}} = \left(\frac{2}{C_L}\right)^{3/2}\left(\frac{C_v^{3/2}}{A_r^{1/2}}\right)\left(\frac{\rho_s}{\rho}\right)^2\left(\frac{t}{c}\right)\left(\frac{W_a g^2}{\rho U^6}\right)^{1/2}$$

Assuming that $W_s/W_a \ll 1$, then $(1 + W_s/W_a)^{3/2} \doteq 1 + (3/2)(W_s/W_a)$.
Using this approximation, the above equation becomes

$$\frac{W_s}{W_a} \doteq \frac{\left(\frac{2}{C_L}\right)^{3/2}\left(\frac{C_v^{3/2}}{A_r^{1/2}}\right)\left(\frac{t}{c}\right)\left(\frac{\rho_s}{\rho}\right)^2\left(\frac{W_a g^2}{\rho U^6}\right)^{1/2}}{1 - \frac{3}{2}\left(\frac{2}{C_L}\right)^{3/2}\left(\frac{C_v^{3/2}}{A_r^{1/2}}\right)\left(\frac{t}{c}\right)\left(\frac{\rho_s}{\rho}\right)^2\left(\frac{W_a g^2}{\rho U^6}\right)^{1/2}} \tag{136}$$

Using standard aerodynamic procedure, C_d is separated into
the profile drag coefficient C_{dp} and the induced drag coefficient
C_{di}, where

$$C_d = C_{dp} + C_{di} \tag{137}$$

Using airfoil data from Abbott and Doenhoff (21) and other sources, Hoerner (18) developed semi-empirical expressions which lead to the following expression for C_{dp} of lifting surfaces with a turbulent boundary layer[1]:

$$C_{dp} = 2C_f \left[1 + 2\frac{t}{c} + 30\left(\frac{t}{c} + \frac{C_L}{5}\right)^4 \right] \qquad (138)$$

where C_f is the turbulent skin friction drag coefficient of a flat plate, and is a function of the Reynolds number and surface roughness.

The standard form of the induced drag coefficient is

$$C_{di} = \frac{C_i C_L^2}{\pi A_r} \qquad (139)$$

where C_i is a coefficient which can be found in most aerodynamic references and is 1.0 for an elliptic spanwise lift distribution, and slightly above one for other distributions.

Substituting Equations 137 to 139 into Equation 131 yields

$$Q = \frac{D_\ell}{W_a} = \left\{ \frac{2C_f}{C_L} \left[1 + 2\frac{t}{c} + 30\left(\frac{t}{c} + \frac{C_L}{5}\right)^4 \right] + \frac{C_i C_L}{\pi A_r} \right\}(1 + \frac{W_s}{W_a}) \qquad (140)$$

---

[1] A check of the expression for C_{dp} with the data of (21) showed close agreement for the NACA 63 to 65-series airfoils with standard roughness and for the uniform-pressure airfoils like the 16-series with standard roughness. All utilized the NACA a = 1.0 uniform pressure meanline. Equation 138 is valid only for cambered lifting surfaces operating at the ideal angle of attack, and assumes that the separation drag on the pressure side is negligible. Also, if the t/c taper ratio is much less than one (i.e., $\phi \ll 1$), an error is introduced in Equation 138 due to insufficient weighting of t/c near the root section; however, this error is small for the values of ϕ considered here.

Strength. The physical relationship regarding structural strength is that the design bending stress f at the root section of a lifting surface must be equal to or greater than the applied bending stress. Therefore,

$$f \geq \frac{M_o t_o/2}{I_o} \tag{141}$$

where M is the bending moment, I is the area moment of inertia of a lifting surface cross section, and the subscript o refers to the root section. Since theory shows that the section modulus $I/(t/2)$ is proportional to $t^2 c$, let C_1 be defined as follows:

$$\frac{I_o}{t_o/2} = C_1 t_o^2 c_o = C_1 \left(\frac{t_o}{c_o}\right)^2 c_o^3 \tag{142}$$

where C_1 is the nondimensional section modulus of the lifting surface cross section, and is constant along the span since the cross-sectional shape is assumed to be constant along the span. Furthermore, let C_2, C_3, and C_4 be defined as follows:

$$\frac{t_o}{c_o} = \left(\frac{t}{c}\right)_o = C_2\left(\frac{t}{c}\right) \tag{143}$$

$$c_o = C_3 c \tag{144}$$

$$M_o = \left(\frac{L-W_s-W_x}{2}\right) C_4\left(\frac{b}{2}\right) = \frac{C_4}{4} bL \left(1 - \frac{W_s+W_x}{L}\right) \tag{145}$$

where the net upward force on the wing semispan is half of $L-W_s-W_x$, and C_4 is the nondimensional distance from the root to the semispan center of pressure in terms of semispan length; also, it is assumed that W_s and W_x are distributed along the span proportional to the lift. Substituting Equations 142 to 145 into Equation 141 gives

$$f \gtreqqless \frac{C_4 \, bL \left(1 - \frac{W_s+W_x}{L}\right)}{4C_1 \, C_2^2 \, C_3^3 \left(\frac{t}{c}\right)^2 c^3} \qquad (146)$$

Substituting Equations 133 and 135, Equation 146 becomes

$$f \gtreqqless \frac{C_4 C_L A_r^2 \rho U^2/2}{4C_1 \, C_2^2 \, C_3^3 \left(\frac{t}{c}\right)^2} \left(1 - \frac{W_s+W_x}{W_a+W_s}\right) \qquad (147)$$

Rewriting,

$$\frac{t}{c} \gtreqqless \sqrt{\frac{C_4}{4C_1 \, C_2^2 \, C_3^3} \left(\frac{\frac{1}{2}\rho U^2}{f}\right) \left(\frac{W_a-W_x}{W_a+W_s}\right) A_r \sqrt{C_L}} \qquad (148)$$

Let K_1 be defined as

$$K_1 = \sqrt{\frac{C_4}{4C_1 \, C_2^2 \, C_3^3} \left(\frac{\frac{1}{2}\rho U^2}{f}\right) \left(\frac{1 - \frac{W_x}{W_a}}{1 + \frac{W_s}{W_a}}\right)} \qquad (149)$$

Then Equation 148 becomes

$$\frac{t}{c} \gtreqqless K_1 \, A_r \sqrt{C_L} \qquad (150)$$

Cavitation. Since cavitation of hydrofoils is not permitted, the critical incipient cavitation number of a hydrofoil must lie above the operating cavitation number σ, which is defined as

$$\sigma = \frac{P-P_v}{\frac{1}{2}\rho U^2} \tag{151}$$

The hydrofoil cross sections which have greatest cavitation resistance are any which are similar to the NACA 16-series airfoils which have a near-uniform pressure distribution. Using the data in (21), the following approximate (linearized) expression for cavitation-free operation was developed for the NACA 16-series forms:

$$\sigma \overset{\geq}{=} 2.45 \frac{t}{c} + 0.56 \, C_L \tag{152}$$

Similar expressions can be developed for other hydrofoil forms. Rearranging Equation 152,

$$\frac{t}{c} \overset{\leq}{=} 0.408 \, \sigma - 0.229 \, C_L \tag{153}$$

Elasticity. The physical relationships dealing with the elasticity of the structure will be presented later. Both flutter and divergence will be considered. Each are known to depend upon the parameter $\frac{1}{2}\rho U^2/E$. No elastic effects occur if E is sufficiently greater than $\frac{1}{2}\rho U^2$.

Viscosity. The nondimensional parameter which was selected to represent viscosity is $U^3/g\nu$. The only place where viscosity enters this problem is in the evaluation of C_f, which is primarily a function of Reynolds number R_e and surface roughness. The form of R_e can be transformed as follows:

$$R_e = \frac{Uc}{\nu} = \left(\frac{U^3}{g\nu}\right) \cdot \left(\frac{cg}{U^2}\right) \tag{154}$$

From Equations 133 and 135,

$$c = \left(\frac{2L}{C_L A_r \rho U^2}\right)^{1/2} = \left(\frac{2}{C_L A_r}\right)^{1/2} \left(\frac{L}{W_a}\right)^{1/2} \left(\frac{W_a^{1/2}}{\rho^{1/2} U}\right) \tag{155}$$

Substituting Equation 155 into Equation 154 gives

$$R_e = \left(\frac{U^3}{g\nu}\right) \left(\frac{W_a g^2}{\rho U^6}\right)^{1/2} \left(\frac{L}{W_a}\right)^{1/2} \left(\frac{2}{C_L A_r}\right)^{1/2} = \left(\frac{W_a}{\rho \nu^2}\right)^{1/2} \left(\frac{L}{W_a}\right)^{1/2} \left(\frac{2}{C_L A_r}\right)^{1/2} \tag{156}$$

Consequently, R_e is seen to be a function of a new parameter $W_a/\rho\nu^2$, and of $(L/W_a)^{1/2}$ and $(2/C_L A_r)^{1/2}$ which result from the solution of the design mission. Notice that an estimate must be made of the last two terms of Equation 156 in order to evaluate the Reynolds number, which in turn is needed together with the surface roughness to provide a value for C_f. The value of C_f may then be used to solve the problem. Once the problem is solved, a better estimate can be made for the last two terms of Equation 156, which in turn provides a better value of C_f. If the original C_f is found to be inaccurate, the problem should be reworked.

Let the parameter r' represent the nondimensional surface roughness of a lifting hydrofoil, and let it be considered as a mission parameter. It is known that C_f can be represented as

$$C_f = C_f(R_e, r') \tag{157}$$

where R_e is given by Equation 156. Consequently, C_f can be estimated at the beginning of a design mission problem by estimating R_e, and then using R_e and r' to evaluate C_f using either Reference (18) or (21). Notice that, in general, the last two terms in the expression for R_e of Equation 156 can be approximated by 1.0 until the design form is better determined. Since C_f varies only slightly with R_e, considerable error in evaluating R_e can be tolerated.

Design equations and the optimization criterion. The design equations, which are used for mapping from the various regions of mission space, are Equations 150 and 153. Other design equations will be added later when elasticity is considered. The optimization criterion is given by Equation 140, and Q is to be minimized.

Mission Parameters and Design Parameters

In view of the design equations, the optimization criterion, Equation 136, and the preceding discussion, the best set of mission parameters appears to be W_x/W_a, $(\rho_s/\rho)^4 W_a g^2/\rho U^6$, $W_a/\rho \nu^2$, U/a, $(P-P_v)/\frac{1}{2}\rho U^2$, ρ_s/ρ, f/E, τ, ϕ, r', and $\frac{1}{2}\rho U^2/f$.

The design parameters are C_L, A_r, t/c, and the cross-sectional form.

Design Equations

The optimization criterion given by Equation 140 shows that t/c must be minimized in order to minimize Q. Consequently, the inequality sign can be removed from Equation 150, which becomes

$$\frac{t}{c} = K_1 \, A_r \, \sqrt{C_L} \tag{158}$$

When cavitation is critical, in the sense that the maximum value of t/c permitted by Equation 153 is less than the optimum value of t/c which results when cavitation is not considered, then t/c should be made as large and as close to the optimum value as possible. Therefore, whenever cavitation is critical, the inequality sign must be removed from Equation 153 so that it becomes

$$\text{(cavitation critical)} \quad \frac{t}{c} = 0.408 \, \sigma - 0.229 \, C_L \tag{159}$$

Viscosity-Limited Optimized Lifting Surfaces

The first subspace of mission space selected for mapping is called Subspace (a) and consists of the point defined as follows:

$$\left.\begin{array}{l} \left(\dfrac{\rho_s}{\rho}\right)^4 \dfrac{W_a g^2}{\rho U^6} = 0 \qquad \dfrac{U}{a} = 0 \qquad \dfrac{f}{E} = 0 \qquad \dfrac{P - P_v}{\frac{1}{2}\rho U^2} = \infty \\[3em] \dfrac{W_x}{W_a} = 0 \qquad \dfrac{\frac{1}{2}\rho U^2}{f} = 0 \qquad \tau = 1.0 \qquad \phi = 1.0 \\[3em] r' = \text{NACA standard roughness, defined in (21)} \\[2em] \dfrac{W_a}{\rho \nu^2} \text{ corresponds to } R_e \doteq 6 \cdot 10^6 \end{array}\right\} \tag{160}$$

Subspace (a) therefore represents a single design mission in which no physical phenomena are significant except viscosity. The strength of

the structural material is infinite, its weight is zero, and it has
no elasticity. There are no Mach number or cavitation effects.
There is no planform or t/c taper and no weight is carried in the
lifting surface. The selected values of r' and $W_a/\rho v^2$ are typical
values for practical wings and hydrofoils.

Since $(\rho_s/\rho)^4 W_a g^2/\rho U^6$ is zero, Equation 136 shows that
$W_s/W_a = 0$. Equation 140 then becomes

$$Q = \frac{D_\ell}{W_a} = \frac{D_\ell}{L} = \frac{2C_f}{C_L}\left[1 + 2\frac{t}{c} + 30\left(\frac{t}{c} + \frac{C_L}{5}\right)^4\right] + \frac{c_i C_L}{\pi A_r} \tag{161}$$

Since Q is to be minimized, Equation 161 shows that A_r should approach
infinity and t/c should approach zero. Since the structural material
is infinitely strong, these limits can be approached without violating any criteria. Consequently, Equation 161 reduces to

$$Q = \frac{2C_f}{C_L} + 0.096\,C_L^3\,C_f \tag{162}$$

Since neither of the two design equations is applicable, the value
of C_L must be obtained from Q as follows:

$$\frac{\partial Q}{\partial C_L} = 0 = -\frac{2C_f}{C_L^2} + 0.288\,C_L^2\,C_f$$

Solving,

$$C_L = 1.62$$

This value of C_L is an optimum point because $\partial^2 Q/\partial C_L^2 > 0$. Also,
notice that the optimum C_L is independent of C_f.

The value of C_d is shown by Equation 161 to be simply $C_L Q$. Using Equation 162,

$$C_d = C_f (2 + 0.096 C_L{}^4) = 0.0100$$

where $C_f = 0.00375$, and was obtained from Reference (21).

The lift-to-drag ratio of the lifting surface is

$$\frac{L}{D}_\ell = \frac{1}{Q} = \frac{C_L}{C_d} = \frac{1.62}{0.0100} = 162$$

The best cross section to use for the lifting surface is probably the $a = 1.0$ meanline described in (21) and set at the ideal angle of attack which is zero. This meanline produces a uniform pressure distribution which would tend to minimize turbulent boundary layer separation (and the associated drag increase), even at the relatively high lift coefficient of 1.62. Since Equation 138 for C_{dp} was developed for airfoils utilizing the $a = 1.0$ meanline, the resulting value of C_d is probably valid, even though the data from which Equation 138 was developed did not extend to such thin sections and high lift coefficients. The camber (i.e., the maximum height of the meanline above a straight line joining the leading and trailing edges) is found from Reference (21) to be 0.0895c. Since the aspect ratio approaches infinity, the lifting surface has a very small chordlength and a long span.

Viscosity- and Strength-Limited Lifting Surfaces

The next subspace of mission space to be mapped is called Subspace (b) and is the four-dimensional subspace defined as follows:

$$\left(\frac{\rho_s}{\rho}\right)^4 \frac{W_a g^2}{\rho U^6} = 0 \qquad \frac{U}{a} = 0 \qquad \frac{f}{E} = 0 \qquad \frac{P-P_v}{\frac{1}{2}\rho U^2} = \infty$$

$$\frac{W_x}{W_a} = \text{variable} \qquad \frac{\frac{1}{2}\rho U^2}{f} = \text{variable} \qquad \tau = \text{variable} \qquad \phi = \text{variable}$$

$$r' = \text{NACA standard roughness}$$

$$\frac{W_a}{\rho \nu^2} \text{ corresponds to } R_e \doteq 6 \cdot 10^6$$

The only design equation is Equation 158, which represents stress. Therefore, all additional equations which are needed to evaluate the design parameters must come from Q, which is given by Equation 140.

Evaluation of K_1 and W_s/W_a. Notice that all parameters in Equations 140 and 158 are either mission parameters or design parameters, except K_1 and W_s/W_a. The nature of K_1 and W_s/W_a must therefore be investigated. Equation 149 shows that K_1 consists of the mission parameters $\frac{1}{2}\rho U^2/f$ and W_x/W_a, the coefficients C_1, C_2, C_3, and C_4, and W_s/W_a.

Equation 142 shows that C_1 is a function of the cross-sectional form. C_2 is seen by Equation 143 to be a function of the t/c taper ratio ϕ. Equations 144 and 145 show that C_3 and C_4 are functions of the planform taper ratio τ. W_s/W_a is seen by Equation 136 to be a function of one mission parameter and four design form parameters; since the value of that mission parameter is zero, W_s/W_a is seen to be zero.

Because τ and ϕ are variable, it is seen that C_2, C_3, and C_4 may be considered as variables as long as they are obtained from feasible values of τ and ϕ.

The evaluation of C_1 is a special problem. C_1 represents the cross-sectional strength of the lifting surface. The value of C_1 is much lower for airplane wings than for hydrofoils, even though both may utilize the same airfoil shape. The reason is that wings are generally far more hollow than hydrofoils which are often solid metal. One factor causing this large difference in cross-sectional strength between wings and hydrofoils is that the size of an airplane wing is a much greater proportion of the vehicle size; consequently, wing weight W_s is so highly critical that the increased thickness of a hollow wing (compared to a solid wing) produces a drag increase which is negligible compared to the drag reduction resulting from reduced wing weight and wing area. A second contributing factor is that the value of $\frac{1}{2}\rho U^2/f$ for airplane wings is normally much lower than for hydrofoils, so that structural stress is relatively less critical thereby permitting the use of more hollow cross sections.

In the generalized design mission, the two factors that contribute to a hollow lifting surface are a relatively high value of $(\rho_s/\rho)^4 W_a g^2/\rho U^6$ and a relatively low value of $\frac{1}{2}\rho U^2/f$. In the present problem, the value of the former parameter is zero, which means that the structural weight is zero. Therefore, according to Equation 140, Q is minimized when t/c is minimized, indicating that the lifting surface should be solid for Subspace (b).

The best cross-sectional shape for a lifting surface is the one having the lowest drag for a given strength. The data of (21) show that the drag coefficients of the 63- to 65-series airfoils, and of airfoils of the 16-series type (for example, the 0010-35 airfoil) are the lowest of those shown, assuming NACA standard roughness and $R_e = 6 \cdot 10^6$. Since the calculated value of C_1 for the 16-series airfoils is 0.087[1], compared to about 0.082 for the 63- to 65-series airfoils, the 16-series airfoils are selected as being best from the combined viewpoint of high strength and low drag. Since their pressure distribution is nearly uniform, the 16-series airfoils are also best for cavitation resistance in case they are used as hydrofoil sections.

Because $(\rho_s/\rho)^4 W_a g^2/\rho U^6$ was selected to be zero, this set of design missions does not apply to practical airplane wing design. In order to broaden this problem so that it will apply to practical wings, C_1 and W_s/W_a are now considered to be variables so that practical values can be selected for them. This means that all of the parameters comprising K_1 are now considered to be variable, so K_1 has been broadened. The broadened K_1 is considered to be a new

---

[1] This value was calculated for an uncambered section. C_1 remains essentially unchanged with camber unless the lower side of the lifting surface is concave. Since the latter seldom occurs in practice, C_1 is assumed to be 0.087. All calculations refer to solid sections.

mission parameter which replaces all of the variable mission parameters that were originally selected to describe this set of design missions. The broadened K_1 is

$$\text{(broadened)} \quad K_1 = \sqrt{\frac{C_4}{4C_1 \, C_2^2 \, C_3^3} \left(\frac{\frac{1}{2}\rho U^2}{f}\right)\left(\frac{1 - W_x/W_a}{1 + W_s/W_a}\right)} \quad (163)$$

The original K_1 was

$$\text{(original)} \quad K_1 = \sqrt{\frac{C_4}{0.348 \, C_2^2 \, C_3^3} \left(\frac{\frac{1}{2}\rho U^2}{f}\right)\left(1 - W_x/W_a\right)} \quad (164)$$

Notice that the original K_1 is a special value of the broadened K_1 where $C_1 = 0.087$ and $W_s/W_a = 0$.

Solution of the set of design missions. It is seen from Equations 140 and 158 that the design parameters which must be evaluated are C_L, A_r, and t/c. Substituting Equation 158 into Equation 140 gives

$$Q = \frac{D_\ell}{W_a} = \left\{\frac{2C_f}{C_L}\left[1 + 2K_1 A_r \sqrt{C_L} + 30\left(K_1 A_r \sqrt{C_L} + \frac{C_L}{5}\right)^4\right] + \frac{C_i C_L}{\pi A_r}\right\}\left(1 + \frac{W_s}{W_a}\right) \quad (165)$$

where Q must be differentiated with respect to A_r and to C_L in order to find their optimum values.

Setting $\partial Q/\delta A_r = 0$ and simplifying, yields

$$\frac{4\pi K_1 C_f A_r^2}{C_i C_L^{3/2}}\left[1 + 60 \, C_L^{3/2}(K_1 A_r + 0.2\sqrt{C_L})^3\right] = 1 \quad (166)$$

Setting $\partial Q/\partial C_L = 0$ and simplifying, gives

$$C_L{}^2 \left[30(K_1 A_r + 0.6\sqrt{C_L})(K_1 A_r + 0.2\sqrt{C_L})^3 + \frac{C_i}{2\pi C_f A_r} \right] - K_1 A_r \sqrt{C_L} = 1 \qquad (167)$$

Equations 166 and 167 are independent and can be solved by an iteration process for the optimum values of C_L and A_r as a function of K_1. Setting $C_f = 0.00375$ (for NACA standard roughness and $R_e = 6 \cdot 10^6$) and $C_i = 1.0$[1], Equations 166 and 167 are solved and the resulting values of C_L, A_r, t/c, C_d, and L/D_f are plotted in Figure 24 as a function of K_1. The value of t/c was obtained from Equation 158, and the values for L/D_f and C_d were obtained from Equation 161 where $L/D_f = 1/Q$ and $C_d = QC_L$.

Results. Figure 24 provides considerable information on optimized lifting surfaces. Most lifting surfaces will have a value of K_1 lying between 0.01 and 0.10. The only significant shortcoming of Figure 24 is that it does not include the effects of variable angle of attack $\Delta\alpha$ and Mach number or cavitation number σ. If the effects of $\Delta\alpha$ and Mach number or σ were included, the thickness-to-chord ratio t/c would be reduced, in general, from the values shown in Figure 24, and all of the design form characteristics would change.

Within the assumptions made, the results of Figure 24 are applicable to high-performance sailplane wings and to high-performance

---

[1] It is assumed that the planform shapes are close to the optimum ellipse. However, differences in planform shape from the optimum increase C_i only a small amount, unless the taper ratio is less than about 0.15 according to (21).

174

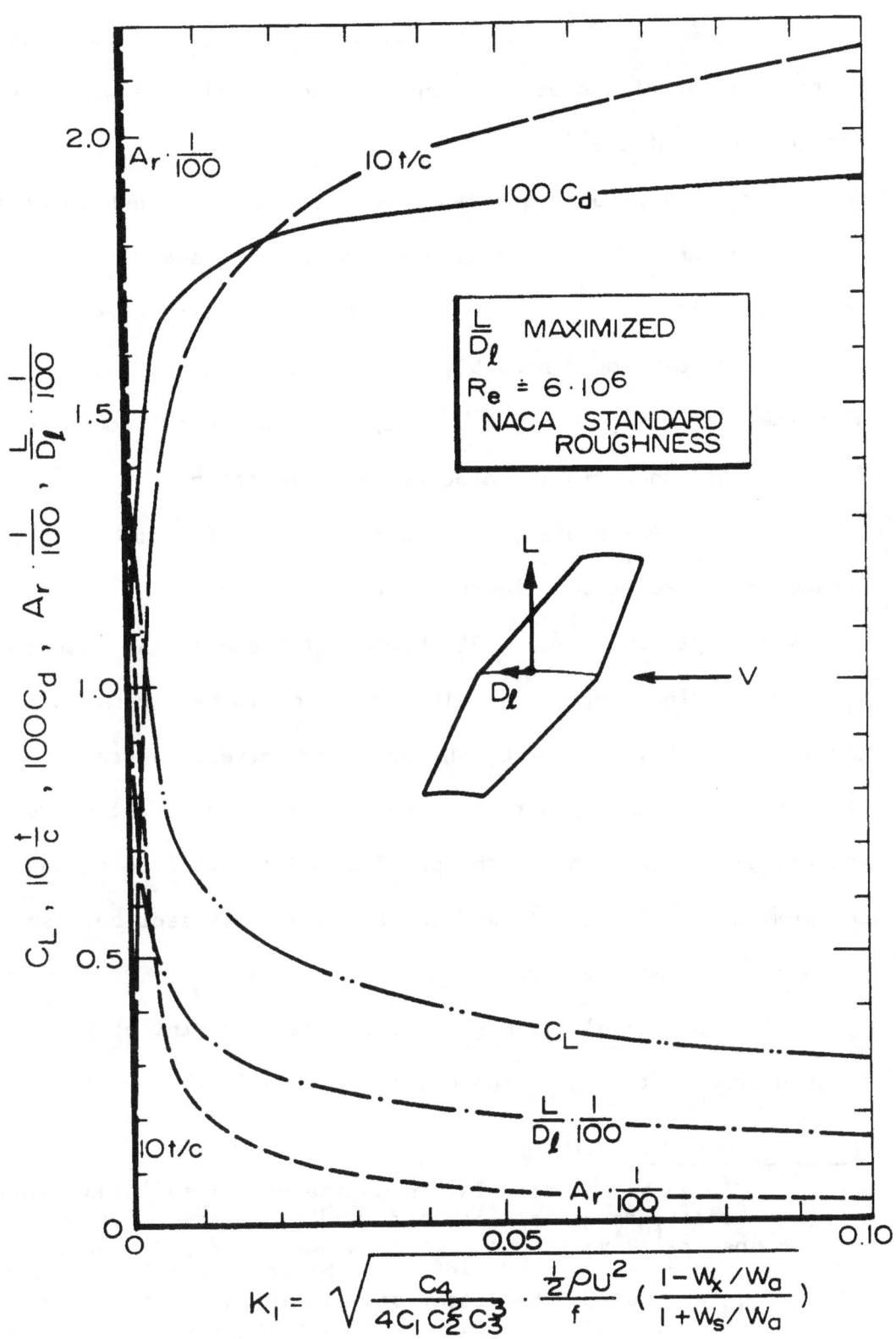

Figure 24 — Viscosity- and strength- limited lifting surfaces

low-speed hydrofoils. For example, a value of $K_1 = 0.01$ corresponds to $A_r = 21$, $C_L = 0.62$, $t/c = 0.173$, and $L/D_f = 36$. These values are characteristic of the best high-performance sailplane wings, except for the value of L/D_f[1].

If the low-speed airplanes were optimized like high-performance sailplanes for flight at a given cruising speed, their wing characteristics would lie closer to the values of Figure 24, and their outer form would probably resemble sailplanes. However, most airplanes are designed for a variety of flight conditions, so factors other than minimum drag are also important.

Optimum value of K_1. Since the values of C_1 to C_4 n the parameter K_1 are to be selected, the question arises as to which values are optimum. Figure 24 shows that the drag is reduced as K_1 reduces. Therefore, the value of C_4 should be minimized and the values of C_1, C_2, and C_3 should be maximized. Since the values of C_3 and C_4 depend upon τ, C_2 depends upon ϕ, and C_1 depends upon the cross-sectional shape, the problem reduces to finding the optimum values of τ and ϕ, and the optimum cross-sectional shape which provides minimum drag.

The optimum planform shape from the viewpoint of minimum induced drag is the one which has a taper ratio that most closely

---

[1] Since the boundary layers on the better sailplane wings may be as much as half laminar, and since their surfaces are generally very smooth, the values of L/D_f can be somewhat higher than 36. However, if the body drag is included, the total L/D for the better sailplanes is reduced to a maximum of around 40. Notice that some changes will appear in Figure 24 in view of the effect of a change in C_f on Equations 166 and 167.

approximates an ellipse. However, to minimize C_4 and to maximize C_3, the taper ratio τ should be as small as possible. On the other hand, to maximize C_2, τ should be as large as possible because, as will be seen later, this permits ϕ to reduce which causes C_2 to increase. Since the optimum values are not highly critical, and in order to simplify the problem, it is assumed that the best value of τ is the intermediate one which approximates an ellipse and thereby permits the induced drag to be minimized.

It can be shown that the value of taper ratio which provides the same mean chordlength and aspect ratio as an elliptic planform is $\tau = 0.570$. Since the cross section and lift coefficient are assumed to be constant along the span, the semispan center of pressure is at the centroid, or 0.454 semispan outward from the root. Consequently, the corresponding optimum value of C_4 is 0.454.

The ratio C_3 of the root chordlength to the average chord-length c of a tapered planform can be shown to be $C_3 = 2/(1 + \tau)$. Consequently, the value of C_3 for the assumed optimum planform is $C_3 = 1.27$.

The optimum value of C_2 is the thickness-to-chord taper which permits the bending stress at all spanwise locations on the lifting surface to be approximately equally critical so that all of the structural material is fully utilized. Setting the maximum bending stress equal to f at all nondimensional distances ζ from the root (where ζ = distance $\div$ b/2), the value $(t/c)_\zeta$ of t/c at ζ for the case of $\tau = 0.570$ and uniform lifting pressure can be shown to be given by

$$\frac{\left(\frac{t}{c}\right)_\zeta}{\left(\frac{t}{c}\right)_o} = \frac{(1-\zeta)(1+1.20\zeta+0.20\zeta^2)}{(1-0.43\zeta)^3} \tag{168}$$

Equation 168 is graphed in Figure 25 which shows that the optimum linearized t/c taper ratio is $\phi = 0.5$. This value is typical for high-performance lifting surfaces. The value of C_2 corresponding to $\phi = 0.5$ is defined as the ratio of (t/c) at the root to (t/c) at the semispan center where $\zeta = 0.500$; therefore, the optimum value of C_2 is 1.33.

The optimum cross sections for solid lifting surfaces were shown earlier to be the NACA 16-series airfoils with a = 1.0 mean-lines. The resulting value of C_1 was 0.087. In the case of a practical airplane wing where wing weight is critical, and the wing is hollow, the optimum value of C_1 can be calculated only if the mission parameter $(\rho_s/\rho)^4 W_a g^2/\rho U^6$ is included in the generalized design mission. The solution of such a generalized design mission would show how hollow a wing should be, and how far C_1^1 lies below the value of 0.087 for a solid wing. Such an analysis can be readily conducted since most of the necessary equations have been developed; however, the analysis is too lengthy to include in this study and would not contribute as much to the illustration of the design procedure as the problems which are considered instead.

---

[1] A typical value of C_1 for a light airplane is 0.0010, or about 1/100 that of a solid section.

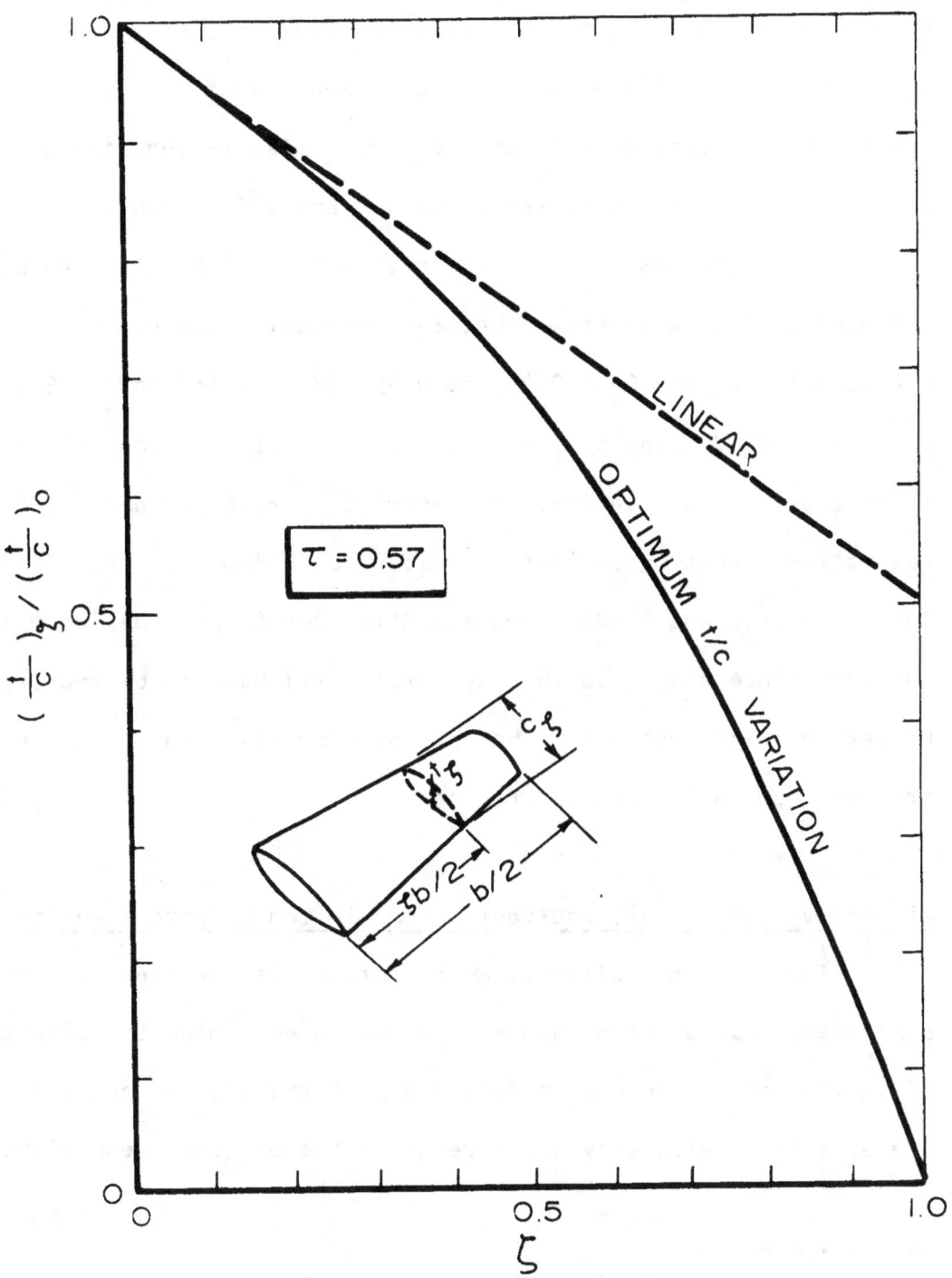

Figure 25 — Optimum spanwise t/c variation

Hydrofoil example. Consider the design of an optimized hydrofoil where $C_1 = 0.087$, $C_2 = 1.33$, $C_3 = 1.27$, and $C_4 = 0.454$. Using Equation 149, $K_1 = 0.610 \sqrt{\frac{1}{2}\rho U^2/f}$ assuming that $W_x/W_a = 0$ and $W_s/W_a \doteq 0$[1]. Assume that two optimized hydrofoils are designed for respective speeds of 30 and 60 knots. Assume that the design stress for a dynamic load factor of 4.0 and a factor of safety of 1.5 is $f = 12,000$ psi. The respective values of K_1 are then 0.023 and 0.047. Figure 24 shows that the respective design form characteristics are $C_L = 0.49$ and 0.39, $t/c = 0.185$ and 0.199, $A_r = 11.5$ and 7.0, and $L/D_f = 26$ and 21. Using Equation 152, the respective critical cavitation numbers σ_{cr} are 0.727 and 0.705. The respective operating cavitation numbers σ, calculated from Equation 151, are 0.825 and 0.206. Consequently, the 60 knot design will cavitate since $\sigma < \sigma_{cr}$, so this hydrofoil will have to be redesigned by taking σ into account. The next problem illustrates the design process when cavitation is critical.

Viscosity-, Strength-, and Cavitation-Limited Hydrofoil Designs

The class of fully-wetted hydrofoils is now considered separately from airplane wings in order to determine the effect of cavitation on the design form and performance, in addition to the effects of viscosity and strength. The original generalized

---

[1] If the hydrofoil had a simple rectangular planform with no t/c taper, then $C_2 = 1.0$ and $C_3 = 1.0$, giving $K_1 = 1.18 \sqrt{\frac{1}{2}\rho U^2/f}$. Notice that the simple form has nearly twice the value of K_1.

design mission and mission criteria are modified to incorporate K_1 as a mission parameter. This design problem is represented by the following subspace of mission space:

$$K_1 = \text{variable} \quad \frac{U}{a} = 0 \quad \frac{f}{E} = 0 \quad \frac{P-P_v}{\frac{1}{2}\rho U^2} = \text{variable}$$

$$r' = \text{NACA standard roughness}$$

$$\frac{W_a}{\rho v^2} \text{ corresponds to } R_e = 6 \cdot 10^6$$

Solution of the equations. The design equations are Equations 158 and 159, and the optimization criterion is given by Equation 161.

Equating t/c in Equations 158 and 159, and solving for A_r,

$$A_r = \frac{0.408 \, \sigma - 0.229 \, C_L}{K_1 \sqrt{C_L}} \tag{169}$$

Substituting Equations 159 and 169 into Equation 161,

$$Q = 2 \, C_f \left[\frac{1}{C_L} + \frac{2(0.408 \, \sigma - 0.229 \, C_L)}{C_L} + \frac{30}{C_L} (0.408 \, \sigma - 0.029 \, C_L)^4 \right]$$

$$+ \frac{K_1 C_i C_L^{3/2}}{\pi(0.408 \, \sigma - 0.229 \, C_L)} \tag{170}$$

Setting $\partial Q / \partial C_L = 0$, and solving for the value of K_1 in terms of the optimum value of C_L gives

$$K_1 = \frac{(\sigma-0.561C_L)^2}{\sigma-0.186C_L}\left[\frac{1+0.816\sigma+0.831(\sigma-0.071C_L)^3(\sigma+0.213C_L)}{156\,C_iC_L^{5/2}}\right] \quad (171)$$

Equation 171 applies only for the case when cavitation affects the design form (i.e., when $\sigma \leq \sigma_{cr}$, where σ_{cr} is the incipient cavitation number of the design form given by Equation 152 with the inequality sign removed). If $\sigma \geq \sigma_{cr}$, then Equations 166 and 167 apply, as in the previous problem.

Results. Equations 171 and the solution to Equations 166 and 167 are graphed in their appropriate regions of Figure 26 which shows σ plotted as a function of K_1 with values of C_L superimposed. As before, C_i has been set equal to 1.0. The use of the parameter K_1 permits Figure 26 to be applicable to hollow hydrofoils, and permits the effect of hydrofoil weight on reducing the bending moment to be included[1]. Notice that a boundary now exists in the graphed section of mission space which separates the strength- and viscosity-limited region where Equations 158 and 161 apply, from the strength- viscosity-, and cavitation-limited region where Equations 158, 159, and 161 (or Equation 171) apply.

The values of design parameters t/c and A_r are plotted as a function of K_1 and σ in Figure 27, and the value of L/D_f

---

[1] Notice that the use of K_1 does not permit the hydrofoil design to be optimized from the viewpoint of its weight, but merely provides a value of K_1 which applies in case the designer wanted to use a hollow hydrofoil and to include the generally small effect of hydrofoil weight on bending moment.

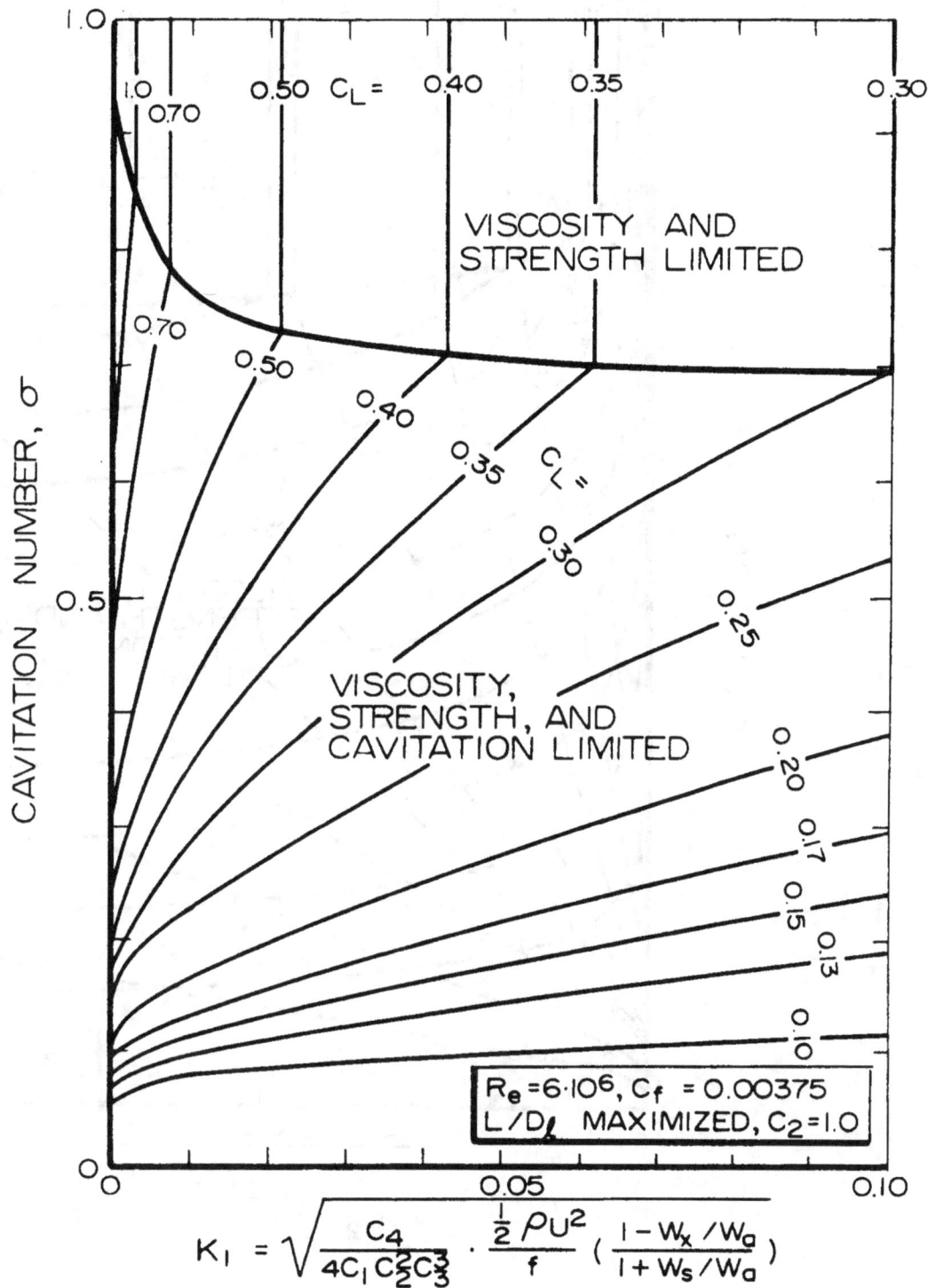

$$K_1 = \sqrt{\frac{C_4}{4C_1 C_2^2 C_3^3}} \cdot \frac{\frac{1}{2}\rho U^2}{f}\left(\frac{1 - W_x/W_a}{1 + W_s/W_a}\right)$$

Figure 26 - Viscosity-, strength-, and cavitation- limited
fully-wetted hydrofoils

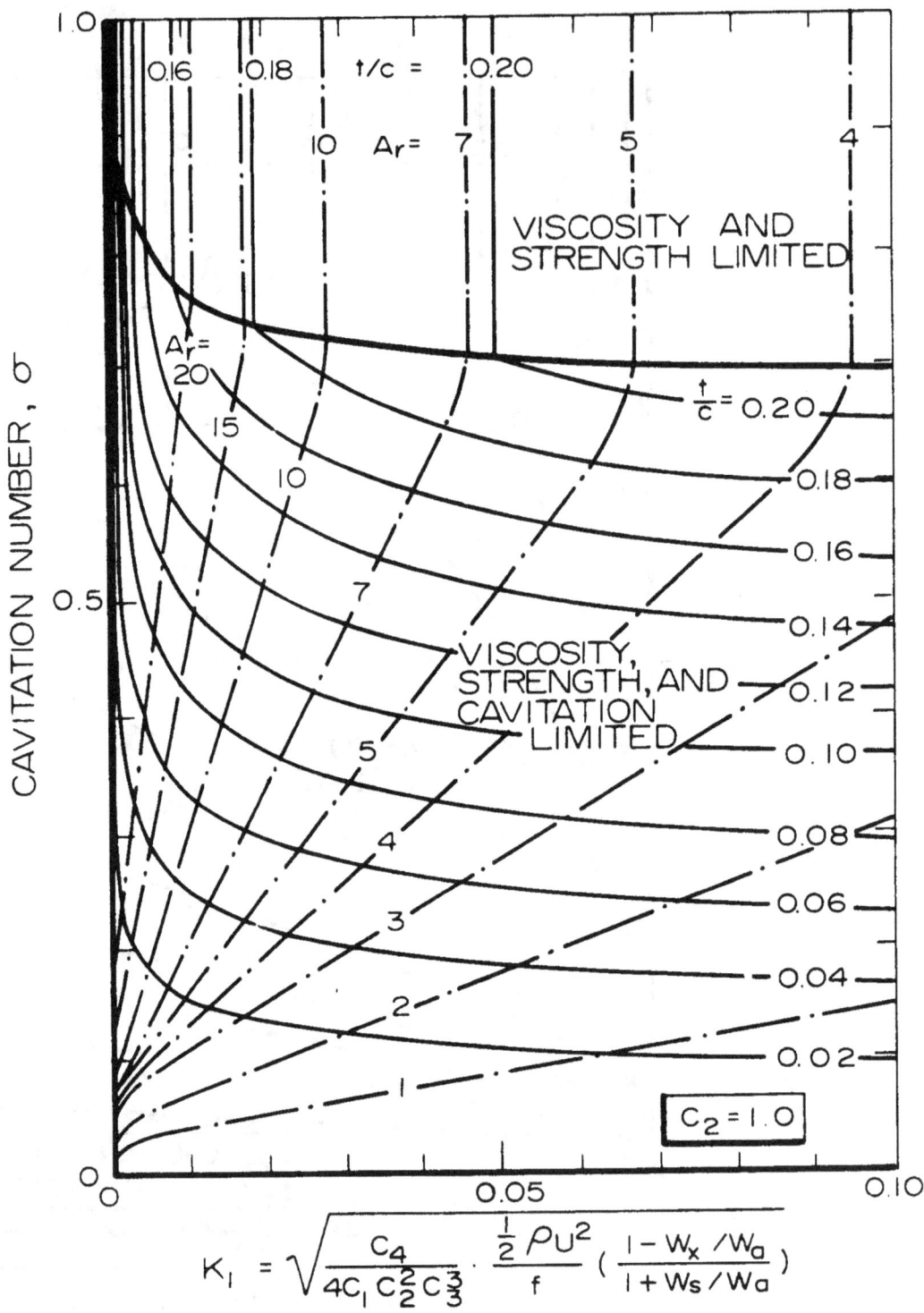

$$K_1 = \sqrt{\frac{C_4}{4C_1 C_2^2 C_3^3}} \cdot \frac{\frac{1}{2}\rho U^2}{f} \left(\frac{1 - W_x/W_a}{1 + W_s/W_a}\right)$$

Figure 27 — Aspect ratio and thickness-to-chord ratio of optimized fully-wetted hydrofoils

(i.e., $1/Q$) is plotted in Figure 28. The values of t/c, A_r, and L/D_f were obtained by using Figure 26 together with Equations 158, 159, and 161. Exactly the same boundary is seen in Figures 26 to 28.

Notice that σ has a strong effect on design form in the region below the boundary in Figures 26 to 28. For example, if the value of K_1 for a hydrofoil is $K_1 = 0.02$, the respective design parameters for two situations where σ is 0.75 and 0.2 are $C_L = 0.51$ and 0.20, $t/c = 0.18$ and 0.04, $A_r = 14$ and 4, and $L/D_f = 27$ and 18.

Some readers may wonder why sweepback was not considered in order to reduce some of the design limitations imposed by cavitation. Appendix D has been added to show that whenever a hydrofoil (designed for a fixed angle of attack) is strength and cavitation limited, that sweepback does not help to change the cavitation limitations and reduce drag.

As a final comment, notice that the results of this problem can be used for propeller blade design. Since Figures 26 to 28 apply to cantilevered hydrofoils, they also apply to propeller blades if the blade twist is low and if both U and σ are properly calculated.

Design of Airplane Wings and Hydrofoils Which are Thickness Limited in Addition to Being Viscosity and Strength Limited

If the design of a lifting surface is thickness limited in addition to being viscosity and strength limited, the optimization criterion is given by Equation 161 (assuming $W_s/W_a = 0$), and the

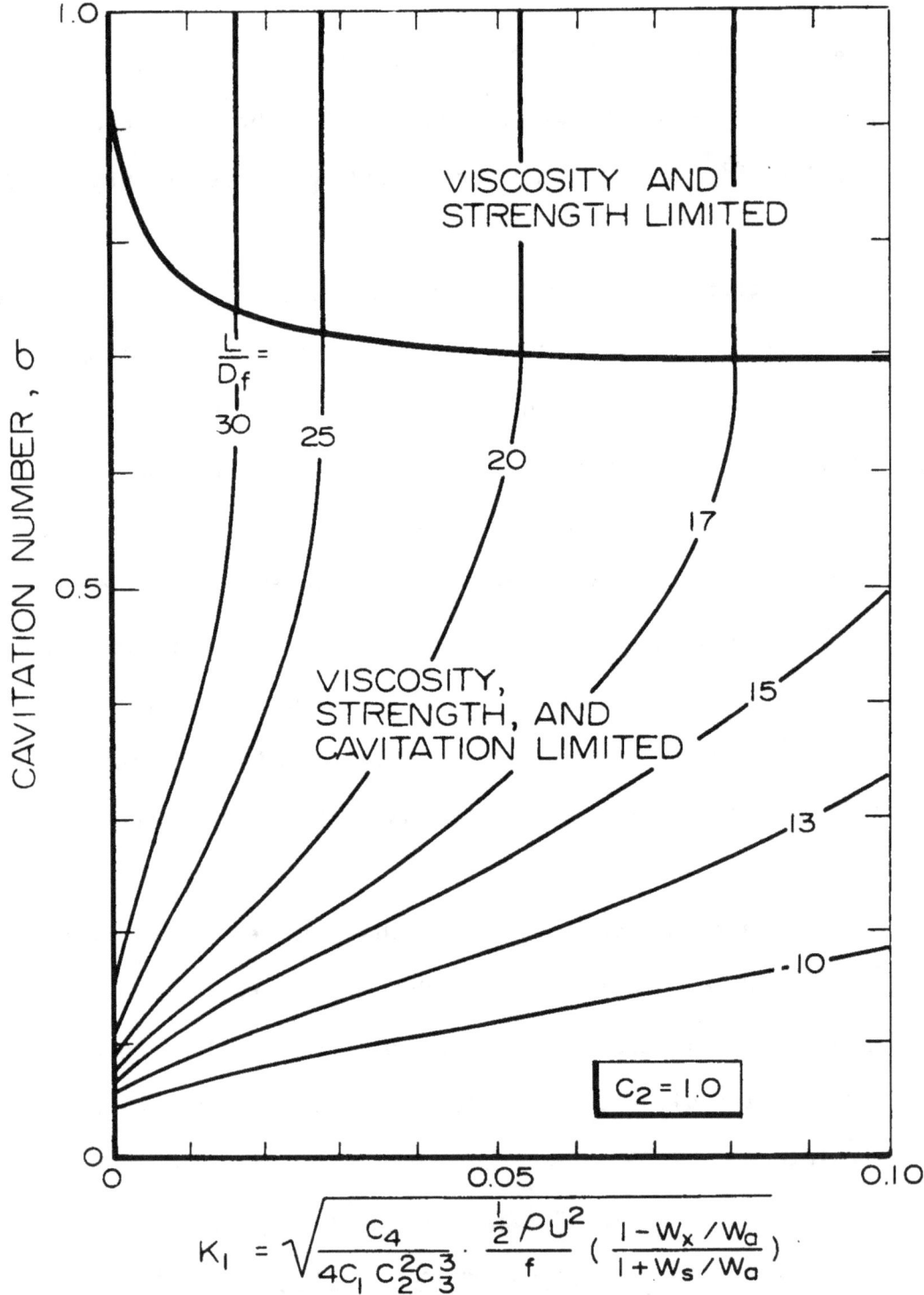

$$K_1 = \sqrt{\frac{c_4}{4c_1\,c_2^2 c_3^3}} \;\; \frac{\frac{1}{2}\rho U^2}{f} \left(\frac{1 - W_x / W_a}{1 + W_s / W_a} \right)$$

Figure 28 — Lift-to-drag ratio of optimized fully-wetted hydrofoils

design equation is Equation 158. If Equation 158 is used in the normal manner, and the resulting t/c is less than the specified maximum, the solution is the same as for the viscosity- and strength-limited problem treated earlier. However, if the value of t/c which results from the use of Equation 158 exceeds the specified maximum value of t/c, then t/c must be fixed at the specified maximum value in Equation 158. In this case, a different solution will result from substituting Equation 158 into Equation 161. Consequently, two regions will appear in mission space.

Alternate procedure. A short cut in solving this problem is to utilize the graphs resulting from the last problem. Even though cavitation has no bearing on this problem, the results of the last problem can still be used. Notice that the value of K_1 is clearly the same in both problems. The only difference between the two problems is that a critical value of σ is selected in the last problem while a critical value of t/c is selected for this one.

Figures 26 to 28 can be used to solve any design mission falling within the scope of this current problem by using the following method: (a) locate the point in Figure 27 which corresponds to the specified maximum value of t/c and the calculated value of K_1 for the given design mission; this point provides the design aspect ratio and also a pseudo-cavitation number; (b) using the pseudo-cavitation number and K_1, find the corresponding values of C_L and L/D_f in Figures 26 and 28.

Of course , this problem could have been easily solved by following the design procedure. This alternate procedure was shown to illustrate that a previous solution can sometimes be applied to solve a different kind of design problem. The use of such an alternate procedure will also save time and effort because additional graphs would otherwise be required.

Example. As an example, consider the case of an airplane wing where the calculated value of K_1 is 0.04 and t/c is limited to 0.10. Figure 27 shows that A_r = 4.8 and σ = 0.39. Using σ and K_1, Figures 26 and 28 show that C_L = 0.27 and L/D_f = 20.

Design of Elasticity-, Cavitation-, Viscosity-, and Strength-Limited Hydrofoils

This final problem of the appendix includes the effects of flutter and divergence which now become important because elasticity is considered. The associated section of mission space is defined as follows:

$$K_1 = \text{variable} \quad \frac{U}{a} = 0 \quad \frac{f}{E} = \text{variable} \quad \frac{P-P_v}{\frac{1}{2}\rho U^2} = \text{variable}$$

$$r' = \text{NACA standard roughness}$$

$$\frac{W_a}{\rho v^2} \text{ corresponds to } R_e = 6 \cdot 10^6$$

Flutter and divergence. The introduction of the modulus of elasticity E presents the possibility of hydrofoil failure by flutter or divergence, although bending moment, which has already been treated, is by far the most important design consideration. Flutter is characterized by an oscillation which beings suddenly when a certain critical speed is reached. It is caused by an interaction of dynamic, elastic, and inertial forces. The oscillation builds up rapidly, and generally results in failure after only a few cycles. Fortunately, no authenticated cases of flutter have been reported on the lifting surfaces of operational hydrofoil craft; however, flutter has been observed in laboratory experiments on swept-back (non-lifting) struts which were very thin and accurately aligned (22). If the thin struts were not accurately aligned, they would instead fail in bending due to the side force produced by the small angle of attack. If the thin struts were not swept back, failure would occur at a somewhat higher critical speed, and would be caused by divergence.

Divergence is a combined dynamic and elastic phenomenon. Beyond a certain critical speed, any small angle of attack produces a critically large hydrodynamic moment about the spanwise elastic axis. This critical value of hydrodynamic moment produces an elastic twist which further increases the angle of attack which in turn produces more hydrodynamic moment and more twist, and so on, until failure occurs in torsion. The entire process is very rapid.

Divergence, when critical, can be eliminated by a sweepback of only a few degrees (23)[1].

Design equation for divergence. Assuming that the hydrofoil is not swept back, divergence, but not flutter, may occur (22). The expressions[2] for the critical dynamic pressure q_d above which divergence occurs on cantilevered hydrofoils having a constant t/c ratio, cantilevered span of b/2, and planform taper ratios of 1.0 and 0.5, according to Reference (23) are:

$$(\tau = 1.0) \quad q_d = \frac{\pi^2 GJ}{ce \, C_{L_\alpha} b^2} \qquad (172)$$

$$(\tau = 0.5) \quad q_d = \frac{10.96 \, GJ_o}{c_o \, e_o \, C_{L_\alpha} b^2} \qquad (173)$$

where GJ = torsional stiffness of the cross section, $J = C_t ct^3$, $C_t = 0.30$ for thin rectangular cross sections (24) which is approximately valid for hydrofoil sections, e = distance from the center of pressure to the elastic axis, $C_{L_\alpha} = \partial C_L / \partial \alpha$, and G = modulus of

[1] When a foil is swept backward, the normal transverse deflection due to lift along the span produces a slight reduction in angle of attack when combined with the sweep angle which opposes the increase in angle of attack due to torsional elasticity.

[2] These expressions for q_d were developed for cantilevered airfoils or hydrofoils assuming that there is no angle of attack deflection in steady flight which would move the center of pressure outward. This assumption is valid for the present problem because the elastic axis is close to the center of lift for all designs.

rigidity = $E/2(1 + \mu) = E/2.6$ where μ = Poisson's ratio = 0.30

for most structural materials. The subscript o refers to the root

section. Substituting the above relationships into Equations 172

and 173, together with $c_o = C_3 c$, $(t/c)_o$ = t/c since t/c is assumed

constant along the span (i.e., $C_2 = 1.0$), $e_o = C_3 e$, $J_o = C_t (t/c)_o^3 c_o^4$

$= C_3^4 J$, $C_3 = 1.0$ for $\tau = 1.0$ and $C_3 = 1.33$ for $\tau = 0.5$, $A_r = b/c$,

the equations become

$$(\tau = 1.0) \quad q_d = \frac{\pi^2 E (0.30)(t/c)^3 c^4}{2.6 c^2 (e/c) C_{L_\alpha} A_r^2 c^2} = \frac{1.14 E (t/c)^3}{(e/c) A_r^2 C_{L_\alpha}} \tag{174}$$

$$(\tau = 0.5) \quad q_d = \frac{(10.96) E (0.30)(t/c)^3 (1.33)^4 c^4}{2.6 (1.33)^2 c^2 (e/c) C_{L_\alpha} A_r^2 c^2} = \frac{2.24 E (t/c)^3}{(e/c) A_r^2 C_{L_\alpha}} \tag{175}$$

Consequently, for the same average chord and aspect ratio, a

tapered hydrofoil with $\tau = 0.5$ has about twice the critical dynamic

pressure of an untapered hydrofoil. This result shows the further

advantage of taper for elastic effects in addition to the previously

shown advantages for structural bending and induced drag.

Letting $q_d = \frac{1}{2}\rho U^2$ so that C_d is minimized (because t/c is

minimized), and substituting C_5 for the numerical constant appearing

in Equations 174 and 175, where C_5 is a function of the taper

ratio τ, the two equations become a single equation which can be

solved for t/c, giving

$$t/c = \left(\frac{e}{c} \cdot \frac{A_r^2 C_{L_\alpha}}{C_5} \cdot \frac{\frac{1}{2}\rho U^2}{E} \right)^{1/3} \tag{176}$$

For aspect ratios around 2.0 and above, standard lifting line theory shows that

$$C_{L_\alpha} \doteq \frac{a_o}{1 + \frac{2}{A_r}} \tag{177}$$

where a_o = value of C_{L_α} for an infinite aspect ratio. Defining K_2 as a new mission parameter where

$$K_2 = \left(\frac{e}{c} \cdot \frac{a_o}{C_5} \cdot \frac{\frac{1}{2}\rho U^2}{E} \right)^{1/3} \tag{178}$$

and substituting Equations 177 and 178 into Equation 176,

$$\frac{t}{c} = \frac{K_2 A_r}{(2+A_r)^{1/3}} \tag{179}$$

Equation 179 is the desired design equation which relates the elastic phenomenon of divergence to the design variables.

Solution of the divergence design problem. The three design equations which must be satisfied when the hydrofoil design is limited by bending stress, cavitation, and divergence, are Equations 158, 159, and 179. Equating t/c of Equation 158 and 179,

$$\frac{t}{c} = K_1 A_r \sqrt{C_L} = \frac{K_2 A_r}{(2+A_r)^{1/3}} \tag{180}$$

Solving for C_L,

$$C_L = \frac{K_2^2}{K_1^2 (2+A_r)^{2/3}} \tag{181}$$

Substituting Equations 180 and 181 into Equation 159,

$$\frac{K_2 A_r}{(2+A_r)^{1/3}} = 0.408 \, \sigma - \frac{0.229 \, K_2^2}{K_1^2 (2+A_r)^{2/3}}$$

Rewriting,

$$\frac{0.229 K_2/K_1^2 + A_r (2+A_r)^{1/3}}{(2+A_r)^{2/3}} = 0.408 \frac{\sigma}{K_2} \qquad (182)$$

The design form can be determined by solving Equation 182 for A_r as a function of the mission parameters σ, K_1, and K_2. t/c and C_L can be obtained from Equations 180 and 181.

The boundary between this region where the design is limited by divergence, strength, viscosity, and cavitation, and the region which is limited by strength, viscosity, and cavitation is obtained by equating t/c of Equation 180 with t/c which is graphed in Figure 27 as a function of σ and K_1. By doing this, Figure 29 was obtained, which is a graph of K_2 versus K_1 where the boundary between the two regions is plotted as a function of cavitation number σ.

Numerical example. It is desired to determine whether divergence is critical. Two hydrofoils are designed for operation near the water surface. They each have a solid 16-series cross-section, no taper, and a constant t/c ratio. Consequently, $C_1 = 0.087$, $C_2 = 1.0$, $C_3 = 1.0$, $C_4 = 0.50$, and $C_5 = 1.14$. They operate at respective speeds of 30 and 90 knots, and are made of a high grade steel where f = 35,000 psi (which includes the load factor and

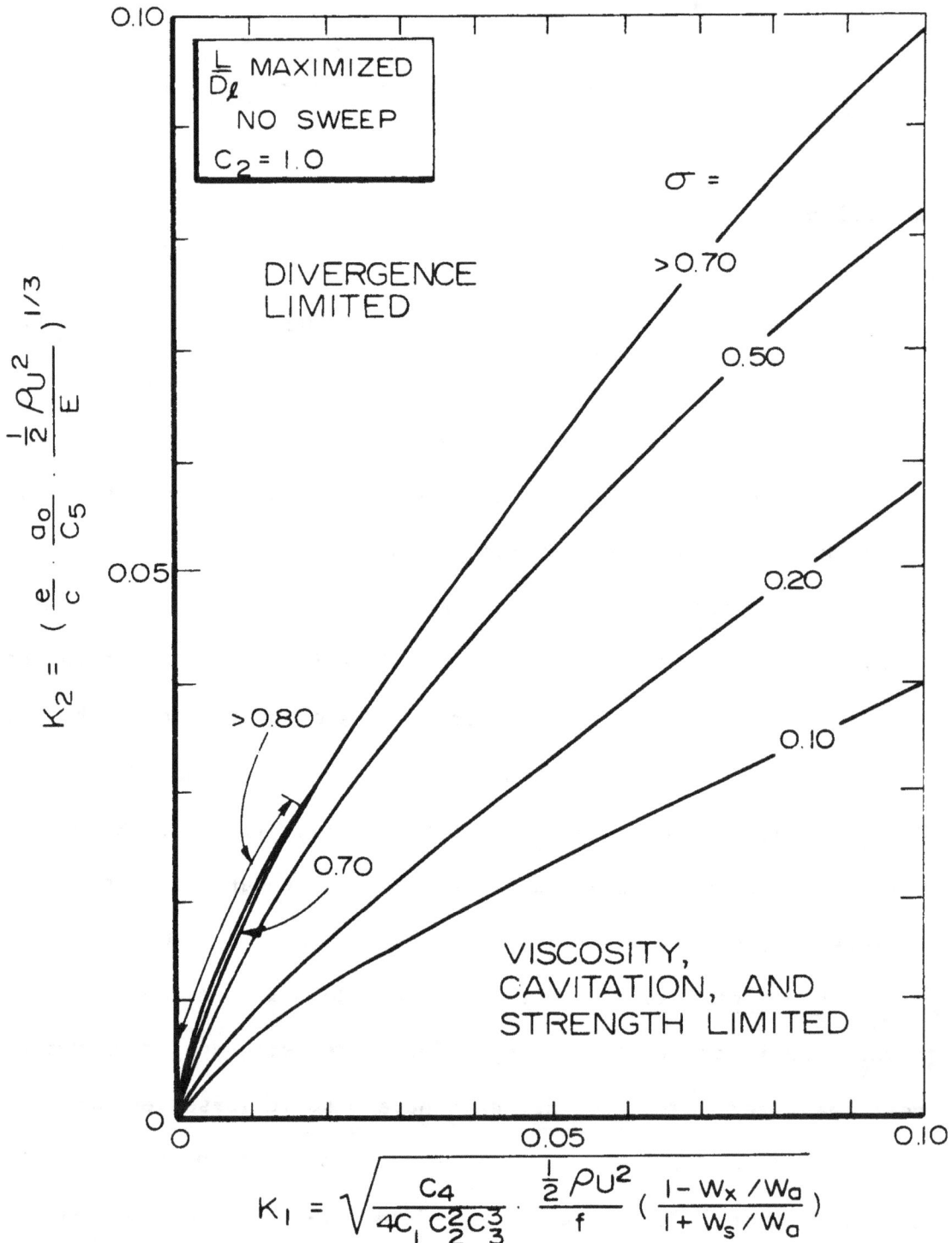

$$K_2 = \left(\frac{e}{c} \cdot \frac{a_0}{C_5} \cdot \frac{\frac{1}{2}\rho U^2}{E}\right)^{1/3}$$

$$K_1 = \sqrt{\frac{C_4}{4C_1\,C_2^2\,C_3^3} \cdot \frac{\frac{1}{2}\rho U^2}{f}\left(\frac{1 - W_x/W_a}{1 + W_s/W_a}\right)}$$

Figure 29 — Divergence limits of optimized fully-wetted hydrofoils

factor of safety) and $E = 30 \cdot 10^6$ psi. The value of e/c for 16-
series cross sections is about 0.23 since the center of pressure
is near the quarter chord and the elastic axis is around the 0.48
chordpoint. It will be assumed that $a_0 = 2\pi$, which is the
theoretical C_{L_α} for thin sections with no boundary layer separation.
Using the above values, the respective mission parameters of the
two hydrofoils are $\sigma = 0.822$ and 0.091, $K_1 = 0.0266$ and 0.0798, and
$K_2 = 0.0091$ and .0190. Figure 29 shows that the boundary for σ
in each case lies above the point determined by K_1 and K_2; therefore,
divergence will not be critical in either case. This result shows
the unlikelihood of divergence failure of optimally-designed lifting
hydrofoils. This result, however, does not eliminate the possibility
of divergence failure for non-optimum lifting hydrofoils or for
(nonlifting) struts.

Design equation for flutter. In case the hydrofoil has
greater than a few degrees of sweep, divergence failure will not
occur, but failure due to flutter might occur. The previous example
showed that divergence failure of unswept, optimized lifting hydrofoils is unlikely. Flutter failure of swept struts is even less
likely since (22) reports that the critical flutter speed of
typical hydrofoils is always greater than the critical divergence
speed.

In analyzing flutter, the theoretical equations which predict
flutter speed for airplane wings do not provide adequate predictions
for hydrofoils. The apparent reason is that the nondimensional

flutter frequency of hydrofoils is an order of magnitude greater
than that of wings. This difference apparently invalidates some
of the assumptions made in the theoretical model used for determining
the values of the coefficients appearing in the unsteady theory.
Fortunately, experimental results on flutter (22) show that a certain
generalization can be made which permits a conservative prediction
of flutter speed. The test results indicate that the nondimensional
flutter speed $U_f/c\omega_\alpha$ lies between 0.5 and 1.0. Selecting 0.5 as a
conservative value,

$$q_f = \tfrac{1}{2}\rho U_f^2 \geq 0.125 \; \rho c^2 \; \omega_\alpha^{\;2} \tag{183}$$

where ω_α is the natural frequency in torsion when submerged, which
from (23) is

$$\omega_\alpha = \frac{\pi}{b} \sqrt{\frac{GJ}{I_\alpha/(b/2)}} \tag{184}$$

where I_α is the mass moment of inertia about the elastic axis.
Substituting the same expressions for GJ as used in conjunction
with Equation 172, and using (22) for $I_\alpha = \rho bc^4(m' + 0.375)/c_\alpha^{\;2}$ which
includes the virtual mass of water for a hydrofoil oscillating about
the elastic axis, Equation 184 becomes

$$\omega_\alpha = 1.4 \sqrt{\frac{c_\alpha^2 \; C_t \, (t/c)^3}{A_r^2 (0.375+m')} \cdot \frac{E}{\rho c^2}} \tag{185}$$

where C_α = torsional mass moment of inertia coefficient = 2π for
solid elliptical forms and $\doteq 2\pi$ for solid 16-series foils,

C_t = torsional stiffness coefficient $\doteq 0.30$ for solid hydrofoil-like forms (24), m' = ratio of hydrofoil mass to the transverse added mass of the fluid $\doteq 0.89\ (t/c)(\rho_s/\rho)C_h$ where ρ_s is the mass density of the structural material, C_h is the ratio of the weight of the given hydrofoil to the weight of an equivalent solid hydrofoil, and the virtual mass of water was calculated as the water mass enclosed by a circular cylinder of diameter c and length b/2. Substituting the value for m' and simplifying, Equation 185 becomes

$$\omega_\alpha \doteq \frac{2.28\ C_\alpha\ \sqrt{C_t}\ (t/c)^{3/2}\sqrt{E/\rho c^2}}{A_r\ \sqrt{1+2.38(t/c)(\rho_s/\rho)C_h}} \tag{186}$$

Substituting Equation 186 into Equation 183, eliminating the inequality, letting $q_f = \tfrac{1}{2}\rho U^2$ in order to minimize C_d, and rearranging, gives

$$\frac{0.65(t/c)^3}{A_r^2\left[1+2.38(t/c)(\rho_s/\rho)C_h\right]} = \frac{1}{C_\alpha^2\ C_t}\cdot\frac{\tfrac{1}{2}\rho U^2}{E} = K_3^2 \tag{187}$$

Equation 187 is the desired design equation for flutter and must be satisifed together with the other two design equations, Equations 158 and 159.

Flutter boundary. The boundary in mission space between the region where the design is flutter limited and the region where the design is cavitation, viscosity, and strength limited is determined by finding t/c and A_r as a function of σ and K_1 using Figure 27, and then substituting these values into Equation 187 to find K_3. Figure 30 shows the resulting boundary plotted as a function of σ in a graph of K_1 versus K_3. Two sets of boundary lines have been drawn, one of which pertains to solid steel hydrofoils and the other to solid aluminum hydrofoils.

Numerical example. Assuming a few degrees of sweepback to eliminate divergence, the same two hydrofoil design missions are selected as those used for the divergence example. Letting $C_\alpha = 2\pi$ and $C_t = 0.30$, the respective values of the mission parameters are $K_3 = 0.00022$ and 0.00066, $\sigma = 0.822$ and 0.091, and $K_1 = 0.0266$ and 0.0798. Since the points determined by K_1 and K_3 in Figure 30 are well below the flutter boundary in both cases, flutter will not occur.

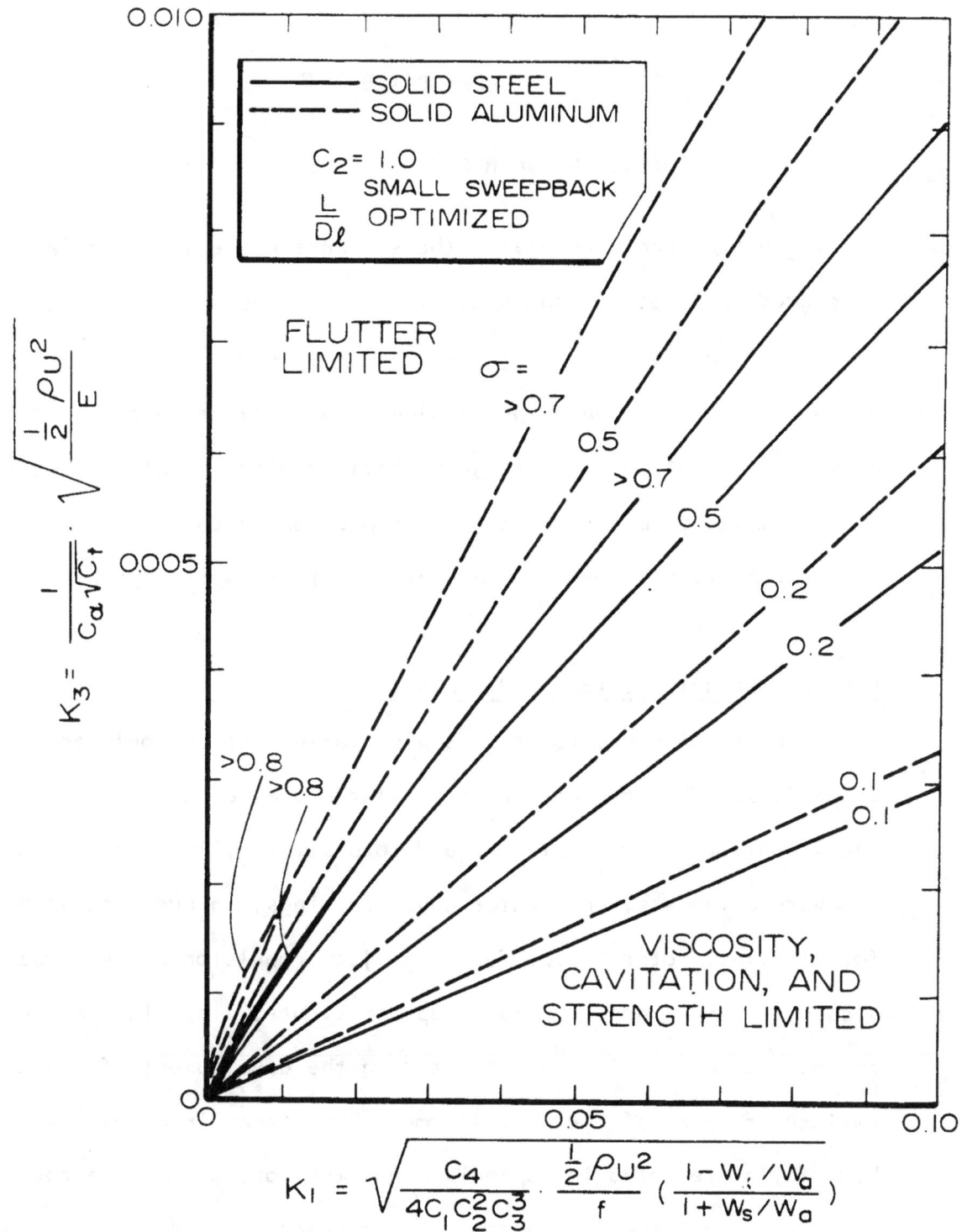

Figure 30 - Flutter limits of optimized fully-wetted hydrofoils

APPENDIX B

THE DESIGN OF HYDROFOIL CROSS SECTIONS

This appendix completes the sequence of design examples presented in Chapter V and Appendix A. The design of hydrofoil cross sections is a subdesign problem of a hydrofoil design problem, which in turn is a subdesign problem of a submerged vehicle design problem. Therefore, the design problem treated in this appendix is twice removed from a complete design problem; because of this remoteness, it has some of the features of a research problem.

General Characteristics of Hydrofoils

Hydrofoils are found in a wide variety of commonly encountered situations. They are used as propeller blades on boats, sailboat keels, ship rudders, submarine and torpedo fins, lifting surfaces of hydrofoil boats, underwater cable fairings, shroud ring stabilizers for missiles, rotor blades for water jet propulsion units, impeller blades in many kinds of pumps, support struts, etc. The many different uses of hydrofoils has resulted in the development of a wide variety of types of hydrofoil forms. The streamlined fully-wetted hydrofoils presented in Appendix A are the most commonly encountered type, and have excellent performance characteristics at speeds up to the beginning of cavitation. Cavitation is characterized by the formation of small cavities filled with water vapor which appear and collapse in the low-pressure region near the hydrofoil surface.

As cavitation increases, there is a correponding increase in the number and degree of such undesirable characteristics aɔ noise, drag, surface pitting, reduction in lift, and unsteady performance. Cavitation can be avoided in certain situations by reducing speed, reducing the hydrofoil thickness or lift coefficient, improving the cross-sectional shape, increasing the free-stream pressure, or by operating closer to the design angle of attack of the hydrofoil.

If cavitation cannot be avoided, an entirely different type of hydrofoil can be utilized which provides steady performance, but has somewhat more drag than the best fully-wetted hydrofoils, and produces more noise. One form is called a supercavitating hydrofoil which is analyzed by Tulin and Burkart (26) and operates with its upper surface entirely immersed in a cavity and with its lower surface fully wetted. Another form is a cavitating, non-lifting strut which is analyzed by Tulin (27) and which is entirely immersed in a cavity, except for the nose section.

A third type of hydrofoil is called a ventilated hydrofoil, various forms of which are described by Lang (28). Ventilated hydrofoils characteristically operate with a steady cavity of noncondensing gas in contact with the surface. At cavitation numbers greater than zero, this type has lower drag than a cavitating hydrofoil, and it operates more quietly. Its use requires a gas source to maintain the cavity.

For the purpose of this appendix it is assumed that a gas source is not available and that the hydrofoils are either fully wetted or else designed for cavitation. Some of the advantages and disadvantages of the various hydrofoil cross sections will become evident later in this appendix.

Specification of the Generalized Design Mission

Many hydrofoil design problems can be reduced to the need for a hydrofoil cross section which provides a certain lift coefficient, sustains a given bending moment, and operates well at a given cavitation number, Reynolds number, etc. A variety of generalized design missions could be considered. The optimization criterion, for example, could be to minimize drag, maximize lift-to-drag ratio, minimize manufacturing cost, minimize drag and weight, etc. The mission criteria could require solid hydrofoil cross sections for various reasons, such as economic, simplicity of fabrication, damage resistance, etc. On the other hand, hollow sections could be required for other equally-valid reasons which pertain for other operating situations. Also, the mission criteria could require noncavitating operation or nonventilating operation. Therefore, a selection must be made of the criteria to be used for the example presented in this appendix.

Selected mission objective and criteria. The selected objective for the generalized design mission is to design hydrofoil cross sections which have minimum drag. The lift coefficient and nondimensional bending moment are to be among the specifications of any given design mission.

The mission criteria are: (a) all hydrofoils must be solid, (b) no ventilation is permitted, (c) the flow is steady, (d) the angle of attack is steady, (e) the operating depth is sufficient to eliminate surface effects, (f) the only critical stress is bending stress, (g) the Froude number is infinite, and (h) the lift coefficient lies between zero and 0.6. The reason for the latter criterion will become apparent later. This permissible range in lift coefficient is acceptable for the vast majority of design applications.

Mission variables. The mission variables are the design stress f of the structural material[1], hydrofoil chordlength c, characteristic surface roughness r, free-stream speed U, free-stream pressure P, fluid viscosity ν, fluid density ρ, fluid vapor pressure P_v, design lift coefficient C_L, and applied bending moment M. In summary, the ten mission variables are f, c, r, U, P, ν, ρ, P_v, C_L, and M. The only nondimensional variable is C_L.

A possible set of mission parameters. The pi theorem predicts six nondimensional parameters, plus C_L. One set of parameters is C_L, M/fc^3, $P/\frac{1}{2}\rho U^2$, $P_v/\frac{1}{2}\rho U^2$, Uc/ν, r/c, and $f/\frac{1}{2}\rho U^2$.

Optimization criterion. Since the hydrofoils are to be designed for minimum drag, the nondimensional optimization criterion is

$$Q = \frac{D/b}{c\frac{1}{2}\rho U^2} = C_d \tag{188}$$

where D/b is the drag per unit spanlength, C_d is the drag coefficient, and Q is to be minimized.

---

[1] The design stress includes the load factor and the factor of safety.

Possible Design Forms

Typical hydrofoil cross sections are sketched in Figure 31. Notice the difference in form between the fully-wetted and the cavitating hydrofoils.

Physical Relationships

The bending stress is considered first, and is obtained by combining Equations 141 and 142. The design bending stress of an arbitrary hydrofoil cross section must be greater than or equal to this bending stress, so

$$f \geqq \frac{M}{C_1 \left(\frac{t}{c}\right)^2 c^3} \tag{189}$$

where c = chordlength, t = hydrofoil thickness, and C_1 is the section modulus coefficient. Nondimensionalizing Equation 189 gives,

$$M' = \frac{M}{fc^3} \leqq C_1 \left(\frac{t}{c}\right)^2 \tag{190}$$

where M' is defined as M/fc^3. Because C_d increases with t/c according to References (18) and (21) for either fully wetted or cavitating hydrofoils, the inequality sign can be removed from Equation 190 in view of the optimization criterion. Therefore,

$$M' = \frac{M}{fc^3} = C_1 \left(\frac{t}{c}\right)^2 \tag{191}$$

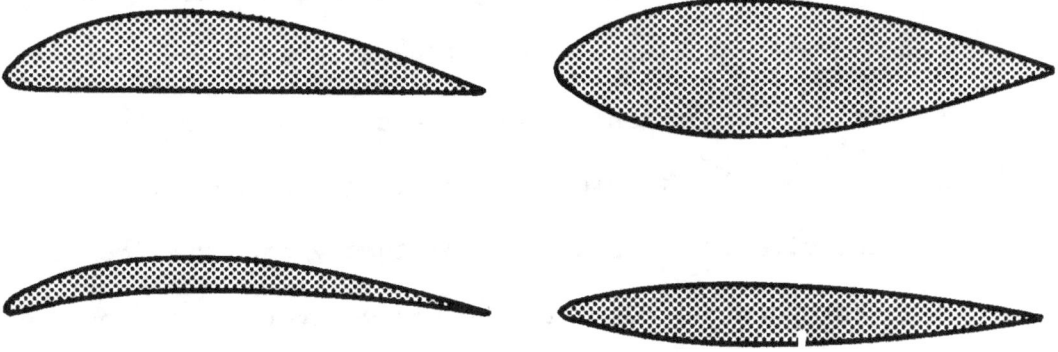

Fully-wetted hydrofoils

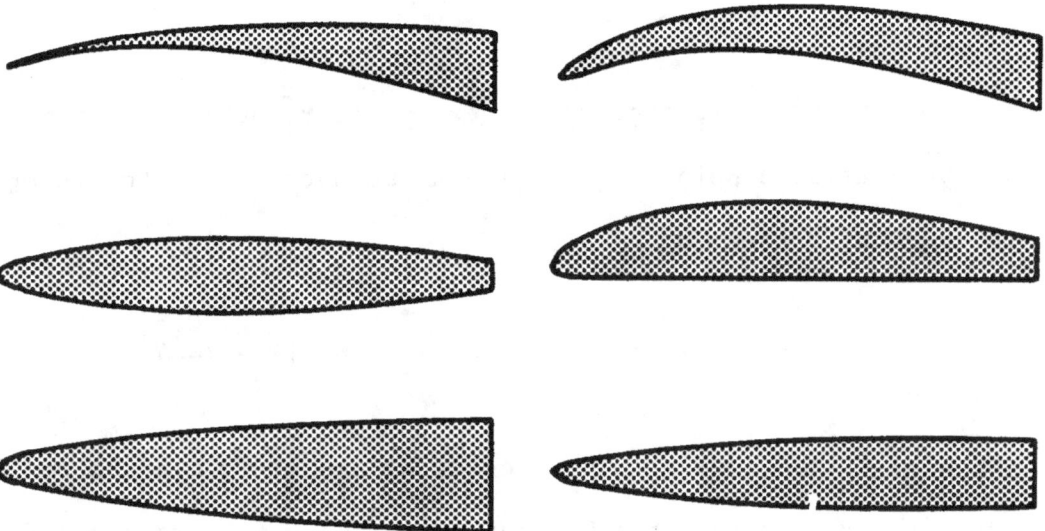

Cavitating hydrofoils

Figure 31 - Typical hydrofoil forms

The viscosity of the operating fluid is considered next. Viscosity affects hydrofoil design because experiments have shown that such drag-producing viscous effects as boundary layer transition and separation can be controlled to a certain extent by hydrofoil form. Consequently, to minimize the drag, the hydrofoil form must be optimized. It is known from theory and experiments that, for a given hydrofoil form, the boundary layer state (and drag coefficient) is a function of the Reynolds number R_e. Consequently, the nondimensional parameter R_e is introduced from the physical viewpoint where

$$R_e = \frac{Uc}{\nu} \tag{192}$$

Cavitation is considered next. Let P_1 be the minimum pressure at some point on a fully-wetted hydrofoil. According to the Bernoulli equation, P_1 is

$$P_1 = P + \tfrac{1}{2}\rho U^2 - \tfrac{1}{2}\rho U_1^2 = P + \tfrac{1}{2}\rho U^2 \left[1 - \left(\frac{U_1}{U}\right)\right]^2 \tag{193}$$

where U_1 is the local fluid velocity at the minimum pressure point. Caviation will occur when P_1 reduces to the vapor pressure of the fluid P_v (assuming no tensile stress in the fluid). The critical (incipient) cavitation number is defined as

$$\sigma_{cr} = \frac{P-P_1}{\tfrac{1}{2}\rho U^2} = \left(\frac{U_1}{U}\right)^2 - 1 \tag{194}$$

where Equation 193 is used to convert the pressures. It is known that σ_{cr} can be expressed functionally as

$$\sigma_{cr} = \sigma_{cr} \left[\frac{t}{c}, \ \overline{t}(x'), \ y_m'(x') \right] \tag{195}$$

where the parameters in the function are $\overline{t}$ = local thickness/t,

x' = x/c = dimensionless distance from the leading edge, and y_m'

= ratio of the local meanline height to the chordlength. Cavitation

will occur whenever $\sigma < \sigma_{cr}$ where σ is the cavitation number which

is defined as

$$\sigma = \frac{P-P_v}{\frac{1}{2}\rho U^2} \tag{196}$$

The roughness of the hydrofoil surface is known to affect

performance. Both drag and lift can be markedly affected by

roughness because roughness can significantly change the boundary

layer state and the flow around the hydrofoil. Experiments have

shown that a nondimensional roughness parameter, for a given

roughness form and distribution, is r' = r/c where r is a charac-

teristic roughness height.

The lift coefficient C_L is already nondimensional, but it

is defined here for convenience as

$$C_L = \frac{L/b}{c \ \frac{1}{2}\rho U^2} \tag{197}$$

Mission and Design Parameters

In view of the list of possible mission parameters and the

physical relationships, the five following mission parameters are

selected: C_L, M', σ, R_e, and r'. Notice that the parameter $\frac{1}{2}\rho U^2/f$

has not been included; the reason is that no relevant physical relationship was found which utilized it. Also, note that P and P_v were combined into the single variable $(P-P_v)$ in the parameter σ.

The design space will consist of whichever parameters are found to best describe the form of a hydrofoil cross section for the section of mission space being considered. Possible design parameters are the nondimensional thickness distribution $\overline{t}(x')$, the nondimensional meanline height distribution $y_m'(x')$, and the thickness-to-chord ratio t/c.

Selection of Subspaces of Mission Space for Mapping

The selection of subspaces of mission space for mapping requires knowledge of the physical phenomena, since the designer must reduce the large number of possible subspaces to those which are the most significant. In hydrofoil design, all of the selected mission parameters are important since any of them could significantly influence the design form. However, some mission parameters are generally more important than others, such as C_L, M' and σ. C_L is important since it represents the primary performance objective of the hydrofoil. M' and σ are selected because bending stress and cavitation considerations are known to be paramount in determining cross-section and thickness in most design problems.

The Reynolds number R_e is also important since Reynolds number determines how the hydrofoil should be formed to best utilize laminar flow, prevent laminar separation, prevent turbulent separation, and minimize skin friction drag. However, R_e is not normally critical in the range $R_e > 10^7$ because the boundary layer is generally fully turbulent, and changes in Reynolds number in this range have only a small effect on hydrofoil form. r' has an effect on hydrofoil performance only when it exceeds certain critical values which depend upon R_e; the roughness can often be kept below these values.

In view of the above discussion, the selected series of mappings will consist of portions of the subspace formed by C_L, M', and σ where r' = 0 and $R_e >> 10^7$. Since the hydrofoil surfaces are smooth and R_e is very high, the value of the skin friction drag coefficient will be very low, but still significant. The following subspaces of the selected subspace will be mapped:

a) C_L = 0, M' = 0, σ variable.

b) C_L = 0, M' variable, σ variable.

c) C_L variable, M' = 0, σ variable.

d) C_L variable, M' variable, σ = 0.

e) C_L variable, M' = 0.0005, σ variable.

f) C_L variable, M' variable, σ variable.

Mapping From Subspace (a) (C_L = 0, M' = 0, σ variable)

Subspace (a) is the simplest of all the subspaces of mission space to map. Since C_L = 0, the corresponding hydrofoil cross section has no camber, and is set at zero angle of attack to minimize drag.

Since $M' = 0$, it has no thickness in view of Equation 191. Therefore, the hydrofoil form is a thin straight line which parallels
the flow.

This hydrofoil form will not cavitate since $\sigma_{cr} = 0$ (see
Equation 194 where $P_1 = P$) and $\sigma_{cr} \leq \sigma$ (because P must always be
greater than or equal to P_v in Equation 196). Consequently, the
selected hydrofoil form is always fully wetted.

According to Equation 138, the drag coefficient is
$C_{dp} = 2C_f$, where C_f is very small because $r' = 0$ and the Reynolds
number is very large. Notice that $C_d = C_{dp}$ always holds in this
appendix because the term C_{di} in Equation 137 is not applicable
since there are no induced drag effects when hydrofoil cross sections
are considered.

Mapping From Subspace (b) ($C_L = 0$, M' variable, σ variable)

This subspace of mission space is more meaningful than the
first one, and relatively important. Since $C_L = 0$, all points in
the two-dimensional mission space of σ versus M' will map into
uncambered hydrofoils, called strut sections. All laminar boundary
layer effects can be disregarded since the boundary layer will be
fully turbulent at $R_e \gg 10^7$. The problem reduces to finding the
minimum drag hydrofoil strut section as a function of σ and M',
where the boundary layer is turbulent and C_f is very small.

Region boundaries. Consider first, the mapping from the line
defined by M' = 0 in Subspace (b). This mapping is exactly that of
Subspace (a), so the design form corresponding to all points along

this line is a zero-thickness fully-wetted straight line.

Next, consider the region where $M' \neq 0$. Strength is now important, so in accordance with Equation 191, the thickness ratio t/c of the associated hydrofoil will increase with M' since $t/c = \sqrt{M'/C_1}$. Cavitation is also important now, since the incipient cavitation number σ_{cr} of a hydrofoil strut is known to increase with thickness ratio, which in turn increases with M'. Consequently, all struts designed for $M' > 0$ will cavitate at $\sigma = 0$, but will not cavitate when $\sigma \geq \sigma_{cr}$. Therefore, it is clear that the graph of ordinate σ versus abscissa M', which represents Subspace (b), will split into two regions where the upper region (i.e., $\sigma \geq \sigma_{cr}$) maps into noncavitating struts and the lower region (i.e., $\sigma \leq \sigma_{cr}$) maps into cavitating struts.

Experimental studies show that the drag coefficient of a cavitating hydrofoil is generally larger than the drag coefficient of an equivalent strength noncavitating hydrofoil; this statement is especially true for $R_e \gg 10^7$ where C_f of a noncavitating hydrofoil is very small. Consequently, the noncavitating forms which are mapped from the upper region (called Region I) will satisfy the minimum drag criterion better than the cavitating forms which are mapped from the lower region (called Region II). Therefore, Region I should be as large as possible, and should extend to the lowest possible values of σ.

The lower boundary of Region I represents the family of struts which, for given values of strength, have the lowest possible incipient cavitation number σ_{cr}. This family of struts is the set which has a near-uniform pressure distribution. Such a family is approximated by the set of ellipses, assuming that the thickness ratios are in the range of conventional hydrofoils, and therefore small. The approximate relationship of σ_{cr} to t/c, for small thickness ratios, is given by Reference (27) as

$$\text{(ellipses)} \quad \sigma_{cr} = 2 \frac{t}{c} \tag{198}$$

Substituting for t/c from Equation 191, and letting $\sigma_{cr} = \sigma$ along the boundary,

$$\text{(Region I to II boundary, } C_L = 0) \quad \sigma = \frac{2}{\sqrt{C_1}} \sqrt{M'} \tag{199}$$

Equation 199 is the desired equation for the boundary between Regions I and II in Subspace (b).

The expression for C_1 is shown by Equation 142 to be

$$C_1 = \frac{I}{\frac{1}{2} t^3 c} \tag{200}$$

Since the value of I for an ellipse is $I = \pi c t^3/64$, the value of C_1 for all ellipses is 0.098.

Mapping of Region I. Region I maps into the family of fully-wetted struts, and its lower boundary maps into noncavitating elliptical forms. The strut forms associated with points inside Region I are now considered, keeping in mind that the optimization criterion is to minimize C_d. It is known that if the thickness distribution of a strut is elliptical, the pressure along the strut surfaces is nearly constant. If the thickness distribution varies from an elliptical shape, a region of increasing pressure and a region of reducing pressure will result. It is known that the local skin friction drag coefficient C_f for turbulent flow increases in a region of reducing pressure and decreases in a region of increasing pressure. Since these effects tend to cancel, it is seen that the net effect on C_f is small and can probably be neglected for this problem. Also, since the local drag is proportional to C_f times the local velocity squared, it can be shown that when the velocity along the surface is nonuniform (due to variations in pressure) the net drag coefficient increases a small amount, assuming that C_f is essentially constant and the turbulent boundary layer does not separate. Therefore, the struts with the lowest friction drag will tend to be the constant-pressure elliptical forms. Since elliptical struts have a relatively blunt trailing edge, a small, sharp-ended cusp-shaped trailing edge should be added to prevent increased drag due to boundary layer separation, particularly at the larger values of t/c. For the purpose of this subspace, however, let us assume that, at $R_e \gg 10^7$, the separation drag of the struts (particularly the thinner ones) is small, so that friction drag is the primary source of drag. Then the elliptic struts become the best solution for all of Region I whether cavitation is critical or not.

The t/c ratio of any ellipse associated with any point in Region I is obtained from Equation 191 by letting $C_1 = 0.098$, giving

$$\text{(ellipse)} \quad \frac{t}{c} = \sqrt{10.2 \, M'} \qquad (201)$$

In view of the assumptions, the above result is valid only for small values of t/c; however, the results should apply without excessive error (particularly if it is understood that a cusp-shaped trailing edge is added in practice) up to values of t/c of around 1/3. The range of $0 \leqq t/c \leqq 1/3$ is seen by Equation 201 to correspond approximately to the range of $0 \leq M' \leq 0.010$.

Letting $x' = x/c$ and $y' = y/c$ be the coordinates of an elliptical strut cross section centered along the positive x-axis with the leading edge at the origin, the equation of the strut can be shown to be

$$(\text{ellipse}) \quad y' = \pm \frac{t}{c} \sqrt{x' - (x')^2} \qquad (202)$$

Substituting Equation 201 for t/c, the mapping solution for Region I becomes

$$(\text{ellipse}) \quad y' = \pm \sqrt{10.2M' \left[x' - (x')^2\right]} \qquad (203)$$

Mapping from Region II. The general form of the cavitating strut family into which Region II maps will be considered first. Let σ_o be the symbol for σ when $C_L = 0$. Then, for small values of σ_o, the cavity drag per unit span of an object which generates a complete two-dimensional cavity increases with cavity size (29) in accordance with the equation,

$$D_c = \frac{\pi}{8} \sigma_o^2 \frac{\rho}{2} U^2 \ell_c = \frac{\pi}{4} \sigma_o \frac{\rho}{2} U^2 t_c \qquad (204)$$

where ℓ_c = cavity length and t_c = cavity thickness. The cavity drag coefficient is defined as

$$C_{dc} = \frac{D_c b}{bc \; \frac{1}{2}\rho U^2} = \frac{\pi}{8} \sigma_o^2 \; \frac{\ell_c}{c} \qquad (205)$$

Assuming that σ_o is small, Reference (29) also shows that the cavity shape is an ellipse where

$$(\text{elliptical cavity}) \quad \frac{t_c}{\ell_c} = \frac{\sigma_o}{2} \qquad (206)$$

which also holds for an elliptical strut if $t = t_c$ and $c = \ell_c$.

Possible candidates for the typical form corresponding to Region II are:

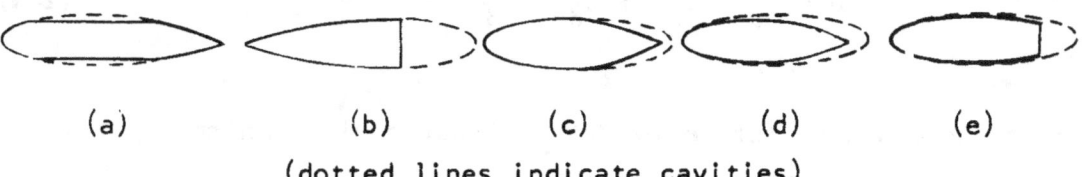

   (a)   (b)   (c)   (d)   (e)

(dotted lines indicate cavities)

Notice that Form (a) can be turned into a fully-wetted strut which would have lower drag by merely filling the cavity to eliminate cavity drag; consequently, it is not a candidate for the cavitating forms of Region II. Form (b) has a larger cavity than necessary, because its trailing edge could be reduced in thickness if its upper and lower surfaces were made more convex until they just began to cavitate everywhere; the resulting form (for equivalent strength) would have essentially no skin friction drag, a relatively small cavity, and would resemble Form (e). Form (c) also has a larger cavity than necessary, for the same reason as Form (b); consequently, the drag of Form (c) would be reduced by a change in form to one which resembles Form (d). However, for a given cavity

drag, the strength of Form (d) can be increased by enlarging its

aft section until it resembles Form (e). Therefore, it is seen

that Form (e), a truncated ellipse, is the best general form

corresponding to Region II, since it provides the least drag for

given values of σ_o and M'.

The optimization criterion for Region II is

$$Q = C_d = C_{dp} = C_{dc} + C_{df} \tag{207}$$

where the profile drag consists only of cavity drag C_{dc} and friction

drag C_{df}. Substituting Equation 205 into Equation 207,

$$Q = C_d = C_{dc} + C_{df} = \frac{\pi}{8} \sigma_o^2 \frac{\ell_c}{c} + C_{df} \tag{208}$$

One design equation for Region II is Equation 191, which is

$$M' = \left(\frac{t}{c}\right)^2 C_1 \tag{209}$$

The other design equation must relate σ_o with the design form,

which is a truncated ellipse that just fits inside an elliptical

cavity defined by Equation 206, or

$$\sigma_o = 2 \frac{t_c}{\ell_c} \tag{210}$$

The form of a truncated elliptical strut can be determined

if t_c/ℓ_c and either c/ℓ_c or t/c are known. The equation for an

ellipse with a semi-axis of length $\ell_c/2$ centered along the positive x-axis with the nose at the origin is

$$y = \pm t_c \sqrt{\frac{x}{\ell_c} - \left(\frac{x}{\ell_c}\right)^2} \tag{211}$$

where y is the local semithickness, and $t_c/2$ is the length of the transverse semi-axis of the elliptical cavity. If the ellipse of Equation 211 is truncated at a point $x = c$, the ratio of the maximum thickness t of the truncated ellipse to its chordlength is:

$$\frac{t}{c} = \begin{cases} \dfrac{t_c}{\ell_c} \cdot \dfrac{\ell_c}{c} & \text{for } \tfrac{1}{2} \leq \dfrac{c}{\ell_c} \leq 1 \\[3ex] 2\dfrac{t_c}{c} \cdot \dfrac{\ell_c}{\ell_c} \sqrt{\dfrac{c}{\ell_c} - \left(\dfrac{c}{\ell_c}\right)^2} & \text{for } 0 \leq \dfrac{c}{\ell_c} \leq \tfrac{1}{2} \end{cases} \tag{212}$$

Substituting Equations 209 and 210 into Equation 212,

$$\frac{t}{c} = \sqrt{\frac{M'}{C_1}} = \begin{cases} \dfrac{\sigma_o}{2} \dfrac{\ell_c}{c} & \text{for } \tfrac{1}{2} \leq \dfrac{c}{\ell_c} \leq 1 \\[3ex] 2\left(\dfrac{\sigma_o}{2}\right) \sqrt{\dfrac{\ell_c}{c} - 1} & \text{for } 0 \leq \dfrac{c}{\ell_c} \leq \tfrac{1}{2} \end{cases} \tag{213}$$

Solving Equation 213 for ℓ_c/c,

$$\frac{\ell_c}{c} = \begin{cases} \dfrac{2\sqrt{M'}}{\sigma_o \sqrt{C_1}} & \text{for } \tfrac{1}{2} \leq \dfrac{c}{\ell_c} \leq 1 \\[3ex] \dfrac{M'}{\sigma_o^2 C_1} + 1 & \text{for } 0 \leq \dfrac{c}{\ell_c} \leq \tfrac{1}{2} \end{cases} \tag{214}$$

Dividing Equation 211 by c, the nondimensional strut semithickness distribution is

$$\frac{y}{c} = \pm \frac{t_c}{\ell_c} \frac{\ell_c}{c} \sqrt{\frac{x}{c} \frac{c}{\ell_c} - \left(\frac{x}{c}\right)^2 \left(\frac{c}{\ell_c}\right)^2} = \pm \frac{t_c}{\ell_c} \sqrt{\frac{x}{c} \frac{\ell_c}{c} - \left(\frac{x}{c}\right)^2} \qquad (215)$$

Substituting Equation 210 into Equation 215, and letting y' = y/c and x' = x/c,

$$y' = \pm \frac{\sigma_o}{2} \sqrt{\frac{\ell_c}{c} x' - (x')^2} \qquad (216)$$

Finally, the strut equation is obtained by substituting Equation 214 into Equation 216,

$$y' = \begin{cases} \pm \dfrac{\sigma_o}{2} \sqrt{\dfrac{2}{\sigma_o} \sqrt{\dfrac{M'}{C_1}} x' - (x')^2} & \left(\begin{array}{l} \text{Region IIa} \\ 10.2 \leq \dfrac{\sigma_o^2}{M'} \leq 40.8 \end{array}\right) \\[4ex] \pm \dfrac{\sigma_o}{2} \sqrt{\left(\dfrac{M'}{C_1 \sigma_o^2} + 1\right) x' - (x')^2} & \left(\begin{array}{l} \text{Region IIb} \\ 0 \leq \dfrac{\sigma_o^2}{M'} \leq 10.2 \end{array}\right) \end{cases} \qquad (217)$$

where Region IIa corresponds to $\frac{1}{2} \leq c/\ell_c \leq 1$, Region IIb corresponds to $0 \leq c/\ell_c \leq \frac{1}{2}$, and the region boundaries were obtained by setting $C_1 = 0.098$ in Equation 214 (recall that $C_1 = 0.098$ for a full ellipse where $c/\ell_c = 1.0$ and notice that C_1 for a semi-ellipse, which corresponds to $c/\ell_c = \frac{1}{2}$, has exactly the same value).

Since two different equations are required to define the strut form in Region II, each equation will be considered to represent a separate family of cavitating struts. The equation for the boundary between Regions IIa and IIb is seen from the region expressions in Equation 217 to be

$$\begin{pmatrix} \text{Region IIa to} \\ \text{IIb boundary,} \\ C_L = 0 \end{pmatrix} \qquad \sigma_o^2 = 10.2\ M' \qquad\qquad (218)$$

The value of C_1 for the two families of truncated ellipses is shown in Figure 32 as a function of the parameter σ_o^2/M'. This relationship was obtained by: (a) integrating over a truncated ellipse to determine the moment of inertia I as a function of c/ℓ_c, (b) calculating the value of C_1 where $C_1 = 2I/t^3c$, and (c) obtaining σ_o^2/M' as a function of c/ℓ_c from Equation 214.

Evaluation of the optimization criterion for Region II forms. The expression for mapping the value of the optimization criterion $Q = C_d$ into Region II of mission space is obtained by substituting Equation 214 into Equation 208 and letting $C_{df} = 0$[1], giving

$$Q = C_{dc} = \begin{cases} \dfrac{\pi}{4} \sigma_o^2 \sqrt{\dfrac{M'}{\sigma_o^2 C_1}} & \begin{pmatrix} \text{Region IIa} \\ C_L = 0 \end{pmatrix} \\[4mm] \dfrac{\pi}{8} \sigma_o^2 \left(\dfrac{M'}{\sigma_o^2 C_1} + 1 \right) & \begin{pmatrix} \text{Region IIb} \\ C_L = 0 \end{pmatrix} \end{cases} \qquad (219)$$

where C_1 is obtained from Figure 32.

The upper expression of Equation 219 is not quite correct, since it does not include the effect of the thrust produced by

---

[1] All forms corresponding to Region II are fully cavitating except the straight line form. There is no friction drag on the fully-cavitating forms.

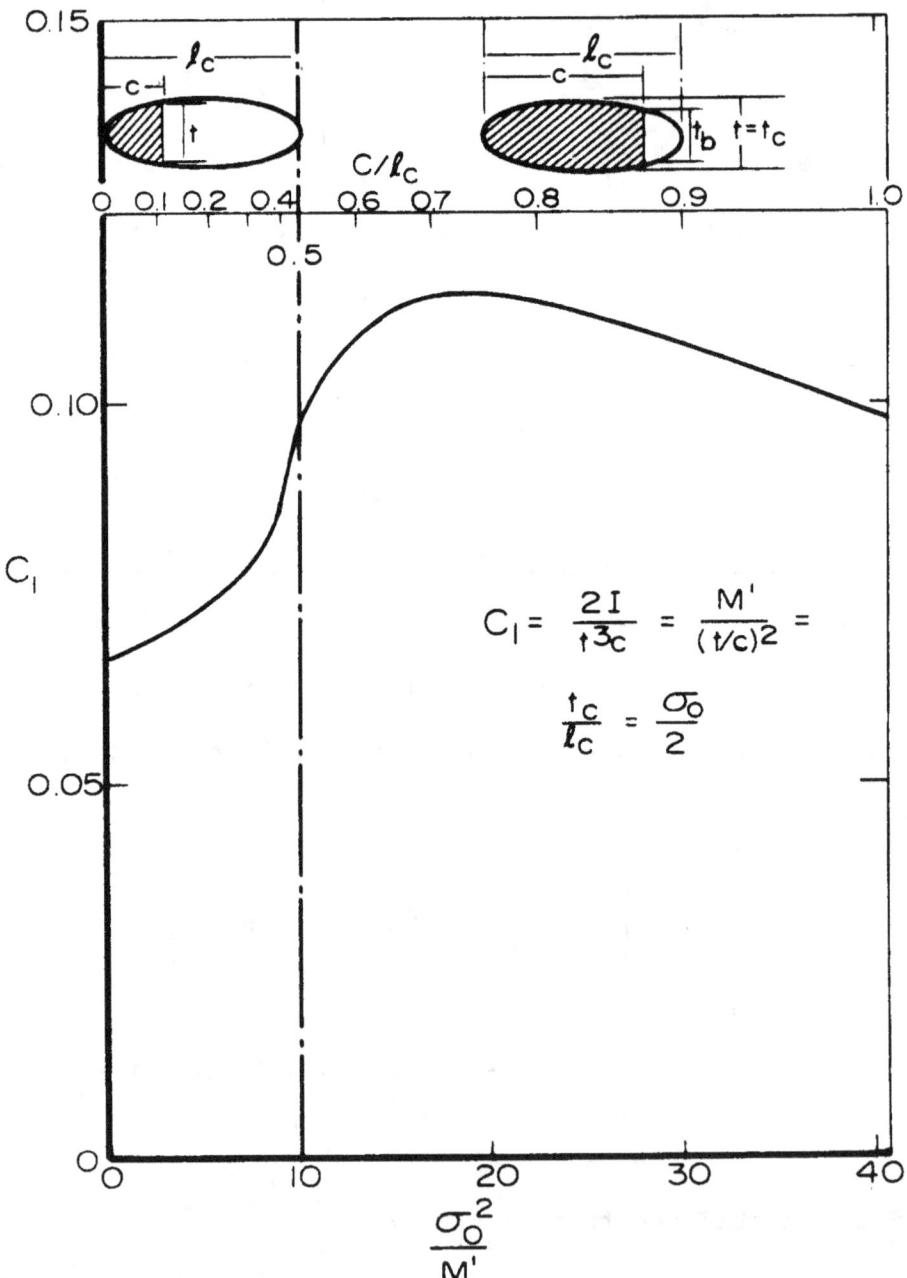

Figure 32 – Section modulus coefficient of truncated elliptical struts

impingement of the reentry jet (which exists at the rear of all cavities) on the strut trailing edge. In theory, if the cavity wall is smooth, and if the trailing edge of the body lying inside the cavity exactly matches the contour of the cavity at its closure point, a thrust will be exerted on the trailing edge which exactly cancels the leading edge drag, leaving zero net cavity drag. In practice, the cavity walls are not smooth, and considerable energy is lost due to turbulence near the cavity collapse point, so all of the theoretical thrust can never be recovered. Also, essentially no thrust is recovered if the trailing edge of the body is so far ahead of the cavity collapse point that the reentry jet does not reach it. In view of the lack of experimental data on the reentry jet effect, it will be assumed that half the theoretical thrust is recovered when the strut almost fills the cavity, and that the effect tapers to zero when the trailing edge of the strut is located more than one-quarter of a cavity length ahead of the cavity collapse point.

Evaluation of the optimization criterion for Region I forms. The value of C_d for fully wetted elliptical struts is a function of the skin friction drag coefficient C_f and the velocity along the strut surface, assuming that pressure drag is negligible. Since both sides of a strut contribute to drag, the drag coefficient of a strut is

$$\text{(Region I, } C_L = 0) \quad C_d = C_{df} = 2C_f\left(\frac{U_1}{U}\right)^2 \qquad (220)$$

where U_1 is the velocity adjacent to the strut surface. The Bernoulli equation is

$$P + \tfrac{1}{2}\rho U^2 = P_1 + \tfrac{1}{2}\rho U_1^2 \tag{221}$$

where P_1 is the pressure along the strut surface. Utilizing Equations 221 and 194, $(U_1/U)^2$ becomes

$$\binom{\text{Region I}}{C_L = 0} \; \left(\frac{U_1}{U}\right)^2 = \frac{\tfrac{1}{2}\rho U_1^2}{\tfrac{1}{2}\rho U^2} = \frac{P+\tfrac{1}{2}\rho U^2 - P_1}{\tfrac{1}{2}\rho U^2} = \frac{P-P_1}{\tfrac{1}{2}\rho U^2} + 1 = \sigma_{cr} + 1 \tag{222}$$

where σ_{cr} is the critical (incipient) cavitation number of an elliptic strut. Substituting Equation 222 into Equation 220, the drag coefficient of forms corresponding to Region I is

$$\binom{\text{Region I}}{C_L = 0} \qquad C_d = C_{df} = 2C_f\,(\sigma_{cr} + 1) \tag{223}$$

which, by substituting Equation 199 and letting $C_1 = 0.098$, becomes

$$\binom{\text{Region I}}{C_L = 0} \quad C_d = C_{df} = 2C_f\left(\frac{2\sqrt{M'}}{\sqrt{0.098}} + 1\right) = 2C_f(6.39\,\sqrt{M'} + 1) \tag{224}$$

Illustration of the mapping result. Figure 33 consists of two graphs of σ versus M' and illustrates the mapping result. Both graphs show the boundaries between Regions I, IIa and IIb (which are described by Equations 199 and 218). Sketches of the corresponding design forms are superimposed on the lower graph at various selected points. The corresponding value of $Q = C_d$ is plotted on the upper graph of σ versus M'. Notice how the value of C_d in Region II increases as the relative trailing edge thickness increases.

222

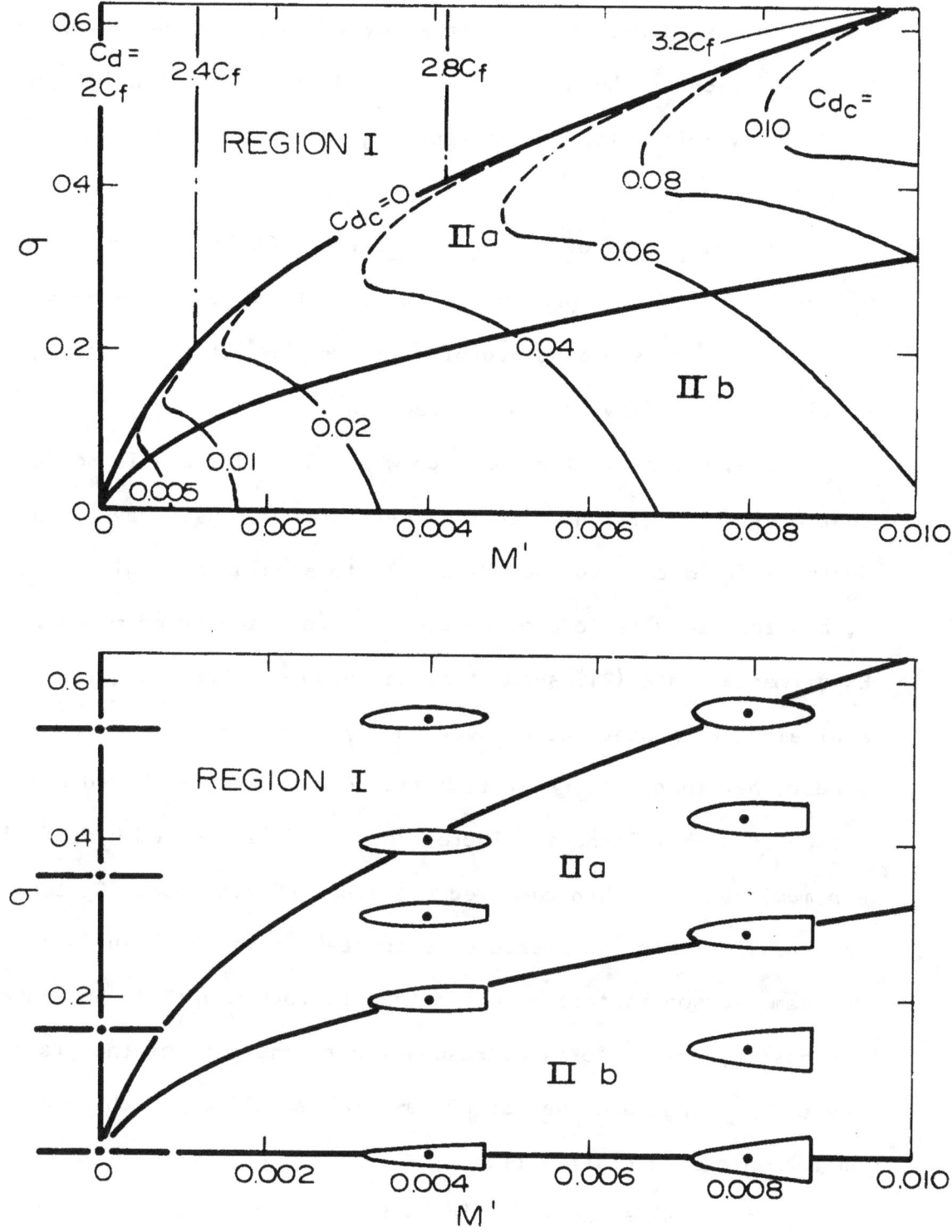

Figure 33 –   Hydrofoil struts and drag coefficients mapped from
                 Subspace (b)

Mapping from Subspace (c) (C_L variable, $M' = 0$, σ variable)

This subspace is represented by a graph of σ versus C_L, where C_L is selected as the abscissa. Since $M' = 0$, Equation 191 shows that all hydrofoil forms corresponding to Subspace (c) will have zero thickness.

Determination of region boundaries. Consider first the hydrofoil forms corresponding to the line $C_L = 0$. Since $M' = 0$ and $C_L = 0$, the corresponding hydrofoil forms are all thin, straight lines parallel to the flow, as in Subspace (a).

Consider next, the subspace where $C_L > 0$ and σ is so large that cavitation will not occur. There are two basic ways of producing lift in this subspace. One is to provide an angle of attack α, and the other is to provide camber (i.e., an arched meanline). Experimental data (21) shows that the drag of airfoils (or equivalently, noncavitating hydrofoils) is minimized when camber is used rather than α to generate lift. As in the uncambered case, C_d increases with thickness. Therefore, the fully-wetted hydrofoil with minimum drag is a thin cambered meanline. Furthermore, C_d tends to be minimized when the pressure is constant along each surface, for the same reason as that given in the previous mapping. Consequently, the best hydrofoil forms corresponding to the part of the graph where σ is large are the set of cambered meanlines which have a uniform pressure distribution.

Now consider possible forms for operation at $\sigma = 0$ and $C_L > 0$. In this case, the free-stream pressure P is the vapor pressure P_v, since $\sigma = \dfrac{P - P_v}{\frac{1}{2}\rho U^2}$. This means that a cavity will form whenever the static pressure on a hydrofoil is less than P. Consequently, the upper side of a cambered meanline will cavitate, and all the lift must be supplied by increased pressure (relative to free-stream pressure) on the lower surface. Therefore, cavitation, and the relatively high drag associated with it, cannot be avoided at $\sigma = 0$. It is now clear that the graph of Subspace (c) will split into two regions, where the upper region, called Region I, will map into fully-wetted hydrofoils, and the lower region, called Region II, will map into cavitating hydrofoils.

The drag of cavitating hydrofoils is larger than the drag of fully-wetted hydrofoils because the wetted skin friction is very low when $r' = 0$ and $R_e \gg 10^7$, while cavity drag is relatively high in general. Therefore, Region I should be made as large as possible. The cambered, uniform pressure meanlines selected for the upper part of Region I are also found best throughout Region I, since they delay cavitation to the lowest possible value of σ; this lowest value of σ is the value $\sigma = \sigma_{cr}$. If σ is reduced slightly below σ_{cr}, the entire upper surface will suddenly cavitate. Therefore, the boundary between Regions I and II will be the line along which $\sigma = \sigma_{cr}$.

The next problem is to determine the incipient cavitation number σ_{cr} of the meanline corresponding to Region I. Expressions for the velocities U_u and U_ℓ along the upper and lower surfaces, respectively, of a fully wetted meanline are

$$U_u = U + u$$

$$U_\ell = U - u$$

(225)

where u is called the circulation velocity.

The Bernoulli equation is

$$P + \tfrac{1}{2}\rho U^2 = P_u + \tfrac{1}{2}\rho U_u^{\,2} = P_\ell + \tfrac{1}{2}\rho U_\ell^{\,2} \tag{226}$$

Substituting Equation 225 into Equation 226, and solving for $P_\ell - P_u$,

$$P_\ell - P_u = \tfrac{1}{2}\rho U^2 \left[\left(1+\tfrac{u}{U}\right)^2 - \left(1-\tfrac{u}{U}\right)^2 \right] = \tfrac{1}{2}\rho U^2 \left(\tfrac{4u}{U}\right) \tag{227}$$

Substituting Equation 227 into Equation 197, gives

$$C_L = \frac{\text{lift}}{bc\ \tfrac{1}{2}\rho U^2} = \frac{(P_\ell - P_u)bc}{bc\ \tfrac{1}{2}\rho U^2} = 4\,\frac{u}{U} \tag{228}$$

The nondimensional upper surface pressure differential is obtained from Equations 225 and 226 as

$$\frac{P - P_u}{\tfrac{1}{2}\rho U^2} = \left(\frac{U+u}{U}\right)^2 - 1 = 2\,\frac{u}{U} + \frac{u^2}{U^2} \tag{229}$$

Notice that the value of Equation 229 just equals the cavitation number σ when $P_u = P_v$, and the entire upper surface is on the verge of cavitating. Therefore, by the definition of σ_{cr}, Equation 229 is the value of σ_{cr}. Using Equations 228 and 229, the equation for the boundary between Regions I and II is

$$\sigma_{cr} = \sigma_{boundary} = 2\,\frac{u}{U} + \left(\frac{u}{U}\right)^2 = \tfrac{1}{2}C_L + \frac{1}{16}\,C_L^{\,2} \doteq \tfrac{1}{2}C_L \tag{230}$$

where the approximation is valid for small C_L.

Mapping from Region I. The design forms corresponding to Region I are the set of constant pressure meanlines. This set is called the set of NACA a = 1.0 meanlines in (21), and is expressed as

$$\left(\begin{array}{l}\text{NACA a = 1.0}\\\text{meanlines}\end{array}\right) \qquad y_m'(x') = y_o'(x') \cdot C_L \qquad\qquad (231)$$

where y_m' is the desired local nondimensional meanline height and y_o' is the local nondimensional height for $C_L = 1.0$. Table 1 contains a partial list of the values of $y_o'(x')$ which were obtained from (21).

TABLE 1

VALUES OF $y_o'(x')$ FOR THE NACA a = 1.0 (UNIFORM PRESSURE) MEANLINE* AT $C_L = 1.0$

x'	0	0.1	0.2	0.3	0.4	0.5
x'	1.0	0.9	0.8	0.7	0.6	0.5
$y_o'(x')$	0	0.0259	0.0398	0.0486	0.0536	0.0552

*Note that the $y_o'(x')$ meanline is symmetrical about $x' = 0.5$, and its maximum height is $0.0552 \ C_L$ at $x' = 0.5$.

Since the drag is assumed to consist solely of turbulent skin friction drag, the expression for C_d is

$$C_d = C_{df} = C_f \left(\frac{U_u}{U}\right)^2 + C_f \left(\frac{U_\ell}{U}\right)^2 \qquad\qquad (232)$$

where U_u and U_ℓ are the velocities along the upper and lower surfaces, respectively, and C_f is the turbulent skin friction drag coefficient of a flat plate for specific values of R_e and r'. Substituting

Equation 225 into Equation 232,

$$C_d = C_f \left(1 + \frac{u}{U}\right)^2 + C_f \left(1 - \frac{u}{U}\right)^2 = 2 C_f \left(1 + \frac{u^2}{U^2}\right) \qquad (233)$$

Substituting Equation 228,

$$C_d = 2 C_f \left(1 + \frac{1}{16} C_L^2\right)$$

Therefore, for values of $C_L \lesssim 0.6$, C_d will be approximately

$$C_d \doteq 2 C_f \qquad (234)$$

This result shows that, for $C_L \lesssim 0.6$, the drag coefficient is essentially independent of design form in Region I.

Mapping from Region II. The general form of the cavitating hydrofoil family will be considered first. Using a dotted line to indicate a cavity, possible candidates for the general form are:

It is clear that Forms (a) and (b) are not candidates for Region II, since their cavities could be filled in, making them fully wetted and lowering their drag; consequently, at their given values of σ, they would not need to cavitate for proper performance, as would the true candidates for Region II. Notice that the cavity size and drag of Form (c) can be reduced if the pressure on the lower side of the cavity-covered portion is reduced by reducing its local camber; this modification can be made only if the lift is increased on the fully-wetted portion in order to keep C_L constant. The lift of the fully-wetted portion can be increased until its upper surface just begins to cavitate uniformly everywhere. The result is a minimum drag form which has the same general shape as Form (d). Consequently, Form (d) is the best general form corresponding to Region II. This form is called a supercavitating hydrofoil.

The mapping of points from the axis $\sigma = 0$ in Region II into supercavitating hydrofoils is considered first. Considerable theoretical information is available on low drag supercavitating hydrofoils operating at $\sigma = 0$. Following the introduction of low drag supercavitating forms derived from linearized theory by Tulin and Burkart (26), Johnson (30) introduced additional low drag forms. The forms which had lowest drag, however, when strength was not considered, were found to be inferior to other forms when practical values of strength were considered. Also, operating depth was found to have some effect on design form. Consequently, Auslaender (31) and (32) extended the linearized theory to include both strength and depth considerations, in order to obtain general expressions for the characteristics of supercavitating hydrofoils at $\sigma = 0$. The general expressions were programmed on an IBM 1620 digital computer to obtain lift, cavity-drag, -shape, and -section modulus of supercavitating hydrofoils composed of the so-called 2-, 3-, and 5-term, and constant-pressure camber configurations, combined with angle of attack and parabolic thickness. It was assumed that the cavities were filled in with metal to provide maximum strength, and that the metal did not quite contact the cavity wall, so that no frictional drag would appear on the upper surface. The results of the computer study showed that the 5-term camber configuration[1] provided the highest lift to drag ratio L/D at $\sigma = 0$ and infinite depth, assuming that strength was not important. In all cases, the configurations were made to satisfy the requirements that the lower surface was fully wetted and that sufficient thickness was added by using angle-of-attack thickness δ or parabolic thickness τ, or both, so that the upper cavity wall would not intersect the lower surface[2]. Auslaender's results further showed that, when reasonable strength requirements were considered, the basic 2-term camber configuration was generally superior, and the constant pressure camber configuration[3] was a close second.

Since (31) and (32) showed that the drag of the 5-term camber was only several percent lower than that of the 2-term camber when strength is unimportant and that the 2-term camber is superior for all practical values of strength, the 2-term camber is selected as the best form corresponding to the axis $\sigma = 0$.

The next problem is to determine the best forms for points in Region II where $\sigma > 0$. Although both nonlinear and linearized theories exist for determining the lower surface shape, cavity shape, and lift and drag coefficients for the case when $\sigma > 0$

---

[1] Called 5-term in view of the number of terms in a certain trigonometric series used in defining the pressure distribution.

[2] The cavity wall passes through the lower surface of the basic 2-, 3-, and 5-term camber configurations.

[3] This camber configuration is designed so that the pressure on the lower surface is uniform.

(Wu, References 34 to 36), the results would require a computer study to determine which form has the lowest drag for a given C_L, σ, and strength.

A relatively simple solution is to linearly add the appropriate NACA a=1.0 uniform pressure meanline to the appropriate two-term supercavitating hydrofoil form (and cavity) designed for $\sigma = 0$ result is a minimum drag hydrofoil form for $\sigma > 0$. Letting σ_{cr} be the incipient cavitation number of an NACA a = 1.0 meanline, and C_{Lo} be the lift coefficient of a 2-term hydrofoil form at $\sigma = 0$, the lift coefficient C_L of the linearized combination is approximately

$$\text{(Region IIe)} \quad C_L = C_{Lo} + 2\,\sigma_{cr} \qquad (235)$$

where C_L is assumed small, and Region II is now called Region IIe for reasons which will be presented later. Notice that the pressure along the upper surface of the linearized form and cavity combination, which shall be called the Region IIe form, is exactly cavity pressure when $\sigma = \sigma_{cr}$. The nondimensional pressure along the lower surface is approximately the nondimensional pressure at $\sigma = 0$ for the two-term hydrofoil designed for $C_L = C_{Lo}$ plus the pressure σ_{cr}. Setting $\sigma = \sigma_{cr}$, Equation 235 becomes

$$\text{(Region IIe)} \quad C_L = C_{Lo} + 2\sigma \qquad (236)$$

which holds for all hydrofoils in Region IIe. Notice that the NACA a = 1.0 meanline is designed for a lift coefficient of $C_L - C_{Lo} = 2\sigma$.

Also, notice that the same basic 2-term hydrofoil form is super-imposed on various NACA a = 1.0 meanlines along any given line paralleling the boundary line $C_L = 2\sigma$.

The Region IIe form is seen to satisfy the necessary boundary conditions for minimum drag, which are: (1) the upper surface pressure is uniform and matches the cavity pressure, (2) the lower surface is fully wetted, and (3) the resulting form has minimum thickness and minimum cavity drag. Furthermore, the Region IIe form is seen to merge into the Region I form at the boundary between Regions I and IIe, since Equation 235 shows that $C_{Lo} = 0$ along the boundary line where $C_L = 2\sigma$. Therefore, the Region IIe form is seen to change smoothly from a supercavitating 2-term form corresponding to $\sigma = 0$ to the NACA a = 1.0 meanlines corresponding to the line $\sigma = C_L/2$.

The design form for Region IIe can now be expressed as a function of C_L and σ. References 31 and 32 show that the lowest-drag 2-term form designed for $\sigma = 0$ is a linear combination of the 2-term camber line denoted by k and the angle of attack thickness distribution[1] denoted by δ where

$$\text{(Region IIe)} \quad k = 0.875\ C_{Lo}, \quad \delta = 0.0787\ C_{Lo} \tag{237}$$

and C_{Lo} is obtained from Equation 236 where

$$C_{Lo} = C_L - 2\sigma \tag{238}$$

─────────────

[1] Added to prevent the cavity wall from intersecting the lower surface.

The nondimensional heights of the upper cavity wall and the lower hydrofoil surface y_u' and y_ℓ', respectively, are

$$\left(\begin{matrix}\text{Region IIe} \\ \sigma = 0\end{matrix}\right) \qquad y_u' = y_1'(x') \cdot k + y_3'(x') \cdot \delta$$

$$y_\ell' = y_2'(x') \cdot k + y_4'(x') \cdot \delta \qquad\qquad (239)$$

where y_1' through y_4' can either be obtained from (31), or from the approximate values obtained from (31) and listed in Table 2. (Also listed in Table 2 is $y_5'(x')$ which relates to the parabolic thickness distribution which will be used later.) Equation 239 is valid only for low values of C_{Lo} because of the assumptions made in the linearized theory; (32) reports negligible error up to $C_{Lo} = 0.2$, but that considerable error may exist for $C_{Lo} > 0.6$. Therefore, the value of C_L in this analysis is limited to a maximum of 0.6.

TABLE 2

APPROXIMATE VALUES OF $y_1'(x')$ THROUGH $y_5'(x')$ FOR THE BASIC
2-TERM CAMBER, δ-THICKNESS, AND PARABOLIC THICKNESS
DISTRIBUTIONS DESIGNED FOR $\sigma = 0$ AND INFINITE DEPTH

x'	0.0	0.05	0.1	0.2	0.4	0.6	0.8	1.0
y_1'	0	0.009	0.017	0.030	0.053	0.073	0.091	0.107
y_2'	0	0.018	0.037	0.071	0.111	0.102	0.038	-0.085
y_3'	0	0.10	0.16	0.25	0.39	0.50	0.59	0.68
y_4'	0	-0.05	-0.10	-0.20	-0.40	-0.60	-0.80	-1.00
y_5'	0	0.22	0.32	0.45	0.63	0.77	0.89	1.00

The shape of the upper and lower surfaces when $\sigma \geqq 0$ is

(Region IIe)
$$y'_u = y'_1(x') \cdot k + y'_3(x') \cdot \delta + y'_o(x') \cdot 2\sigma$$

$$y'_\ell = y'_2(x') \cdot k + y'_4(x') \cdot \delta + y'_o(x') \cdot 2\sigma$$

(240)

where $y'_o(x')$ is the NACA a = 1.0 constant pressure meanline for $C_L = 1.0$, and the value of the meanline designed for C_L is

(NACA a = 1.0 meanline) $\left(C_L\right)_{a=1.0} = C_L - C_{Lo} = 2\sigma$ (241)

Value of the optimization criterion for Region IIe.

The value of the optimization criterion, $Q = C_d$, for Region IIe is $C_d = C_{dc} + C_{df}$. The cavity drag C_{dc} is a rather complicated function of: the cavity drag at $\sigma = 0$ (called C_{do}), hydrofoil t/c, hydrofoil form, and cavitation number σ. Linearized theory was applied by Fabula (37) and (38) to determine C_{dc} as a function of σ for cavitating wedges, special vented hydrofoil struts having $\overset{\centerdot}{C}_{do} = 0$, parabolas, and parabolas with split flaps. The following empirical expression was developed which matches the graphical results of (37) and (38) to within several percent for hydrofoil-like forms:

(Region II) $C_{dc} = C_{do} + \dfrac{\dfrac{\pi}{4}\left(\sigma\,\dfrac{t}{c}\right)^2}{\sigma\,\dfrac{t}{c} + 1.5\,C_{do}}$ (242)

It can be shown that Equation 242 is approximated by $C_{dc} \doteq (\pi/4)(\sigma t/c)$ for the special case when $\sigma\,t/c > 4C_{do}$; the error is an underestimate of less than four percent.

The value of C_{do}, obtained from (31) and Equation 237, is

$$\text{(Region IIe)} \quad C_{do} = (0.319k + 1.25\delta)^2 = 0.142C_{Lo}^2 \qquad (243)$$

Substituting Equation 243 into Equation 242 gives

$$\text{(Region IIe)} \quad C_{dc} = 0.142C_{Lo}^2 + \frac{\frac{\pi}{4}\left(\sigma\frac{t}{c}\right)^2}{\sigma\frac{t}{c} + 0.213\,C_{Lo}^2} \qquad (244)$$

where t/c is the thickness-to-chord ratio of the combined form and cavity (up to the trailing edge). The value of t/c is obtained from Table 2 and Equation 237 as

$$\text{(Region IIe)} \quad \frac{t}{c} = 0.192k + 1.68\,\delta = 0.300C_{Lo} \qquad (245)$$

The friction drag coefficient C_{df} is approximately $C_f \cdot \left(\frac{U_\ell}{U}\right)^2$, where U_ℓ is the average velocity along the lower surface. The value of U_ℓ is $U - u_o - u$, where u_o and u are small relative to U, u_o is the velocity reduction at $\sigma = 0$ due to C_{Lo}, and u is the velocity reduction on the lower surface due to the lift coefficient $\left(C_L\right)_{a=1.0}$ of the NACA meanline. The latter is found from Equations 228 and 241 to be

$$u = \frac{U}{4}\left(C_L\right)_{a=1.0} = \frac{U\sigma}{2} \qquad (246)$$

The value of u_o is obtained from Equation 226 and the approximate relationship for C_{Lo} where

$$C_{Lo} = \frac{P_\ell - P}{\frac{1}{2}\rho U^2} = 1 - \left(\frac{U_\ell}{U}\right)^2 = 1 - \left(\frac{U-u_o}{U}\right)^2 = \frac{2u_o}{U} \qquad (247)$$

where the lower surface velocity, for this case where no NACA camber
is used, is $U - u_o$. Combining Equations 246 and 247, the total lower
surface velocity is

$$\text{(Region II)} \quad \frac{U_\ell}{U} = 1 - \frac{u_o}{U} - \frac{u}{U} \doteq 1 - \frac{C_{Lo}}{2} - \frac{\sigma}{2} \quad (248)$$

Substituting Equation 248 into the expression for C_{df},

$$C_{df} = C_f \left(\frac{U_\ell}{U}\right)^2 \doteq C_f \left(1 - \frac{C_{Lo}}{2} - \frac{\sigma}{2}\right)^2 \quad (249)$$

Substituting in Equation 238,

$$\text{(Region II)} \quad C_{df} = C_f \left(1 - \frac{C_L}{2} + \frac{\sigma}{2}\right)^2 \quad (250)$$

The net drag coefficient C_d obtained by adding Equations 244
and 250, and using Equation 245, is

$$\text{(Region IIe)} \quad C_d = 0.142 C_{Lo}^2 + \frac{0.236\sigma^2 C_{Lo}}{\sigma + 0.71 C_{Lo}} + C_f \left(1 - \frac{C_L}{2} + \frac{\sigma}{2}\right)^2 \quad (251)$$

which, in terms of C_L and σ obtained from Equation 238, becomes

$$\text{(Region IIe)} \quad C_\alpha = 0.142(C_L - 2\sigma)^2 + \frac{0.236\sigma^2(C_L - 2\sigma)}{\sigma + 0.71(C_L - 2\sigma)} + C_f \left(1 - \frac{C_L}{2} + \frac{\sigma}{2}\right)^2 \quad (252)$$

Illustration of the mapping. Some of the hydrofoil forms
corresponding to Subspace (c) of mission space are shown superimposed
on the lower graph of Figure 34 together with the boundary line
between Regions I and IIe. The values of C_d corresponding to
Subspace (c) are plotted in the upper graph. Only the cavity drag

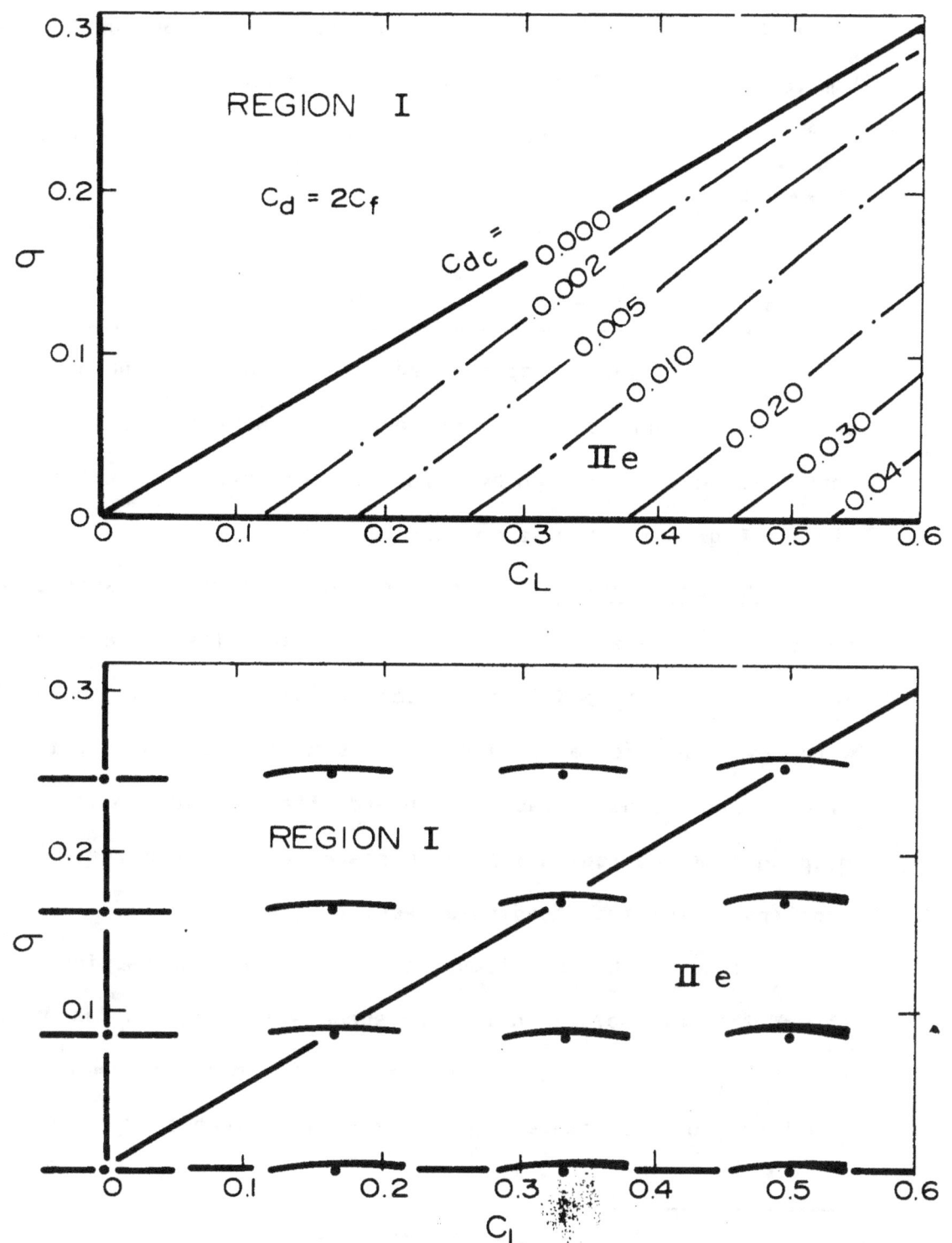

Figure 34 — Hydrofoil forms and drag coefficients mapped from Subspace (c)

is plotted in Region IIe since C_f is negligible relative to C_{d_c} when $R_e \gg 10^7$. Values of C_L are plotted only up to 0.60 due to the limitations of the linearized theory. The most practical range of C_L for supercavitating hydrofoils is around 0.20, so the coverage is adequate[1].

Mapping from Subspace (d) (C_L variable, M' variable, $\sigma = 0$)

Subspace (d) is represented by a graph of ordinate M' versus abscissa C_L, where $\sigma = 0$. A new aspect of Subspace (d) is that it maps into only cavitating hydrofoils; therefore, the entire section lies in Region II of mission space.

Design equations. Before beginning to map, it should be noticed that the axes $C_L = 0$ and M' = 0 have already been mapped. The axis $C_L = 0$ mapped into the set of fully cavitating parabolas with a varying t/c ratio, and the axis M' = 0 mapped into the set of minimum-thickness, two-term supercavitating hydrofoils. Any mapping from the region inside Subspace (d) should merge into these mappings of the two coordinate axes.

Since $\sigma = 0$, the discussion in the previous mapping problem showed that all points in this subspace will map into design forms consisting of different combinations of the two-term camber configuration represented by k, parabolic thickness represented

[1] It is interesting to note that $C_L = 1.0$ is the maximum possible lift coefficient of a supercavitating hydrofoil with steady flow at $\sigma = 0 = (P-P_v)/\frac{1}{2}\rho U^2$ since all of the lift must be generated by positive pressure on the lower surface, and the maximum possible pressure is $\frac{1}{2}\rho U^2$ above vapor pressure.

by τ, and angle-of-attack thickness represented by δ. The parameter δ provides a wedge-like thickness increase. Recall that the forms corresponding to $\sigma = 0$ and $M' = 0$ were represented solely by k and δ. The current problem is to determine the values of k, τ, and δ which provide the hydrofoil form with the lowest C_d as a function of C_L and M'.

The design equation for C_{Lo} (i.e., the value of C_L at $\sigma = 0$) for the 2-term camber is obtained from (31) as

$$\left(\begin{array}{c} \text{Region II} \\ \sigma = 0 \end{array}\right) \qquad C_{Lo} = k + \frac{\pi}{2} \delta \qquad (253)$$

The second design equation should relate M' to the design form parameters k, τ, and δ. Reference (32) does not express M' in equation form, but instead presents a series of graphs of M' (called $\overline{z}$ in Reference 32) shown as a function of k, τ, and δ. By studying these graphs, the following semi-empirical equation[1] for M' was obtained:

$$\left(\begin{array}{c} \text{Region II} \\ \sigma = 0 \end{array}\right) \quad M' = 0.0012(k - \delta - 4\tau)^2 + (0.350\delta + 0.500\tau)^2 \qquad (254)$$

Equations 253 and 254 are the desired equations. These equations can be reduced to one by substituting Equation 253 into

---

[1] Equation 254 is accurate to within a few percent; furthermore, it reduces to the correct value of $M' = 0.269\tau^2 = 0.067(t/c)^2$ for a parabolic strut when $k = \delta = 0$ (i.e., when $C_{Lo} = 0$), and to the value of $M' = 0.122\delta^2 = 0.043(t/c)^2$ for the δ-thickness distribution when $k = \tau = 0$ (the latter value of M' is close to that of a wedge, which is $0.042 \left[t/c\right]^2$).

Equation 254, which gives

$$\left(\begin{matrix} \text{Region II} \\ \sigma = 0 \end{matrix}\right) \quad M' = 0.0012(C_{Lo} - 2.57\delta - 4\tau)^2 + (0.350\delta + 0.500\tau)^2 \quad (255)$$

Optimization criterion. The expression for C_{do} from (31) is substituted into the equation for the optimization criterion to give

$$Q = C_d = C_{do} + C_{df} = \left[0.319k + 1.25(\tau+\delta)\right]^2 + C_{df} \quad (256)$$

Substituting Equation 253 into Equation 256, gives

$$\left(\begin{matrix} \text{Region II} \\ \sigma = 0 \end{matrix}\right) \quad Q = C_{do} + C_{df} = \left[0.319C_{Lo} + 1.25\tau + 0.75\delta\right]^2 + C_{df} \quad (257)$$

Solution of the mapping relations. The expression for Q must be used to solve the problem since two form parameters exist, τ and δ, and only one design equation remains. An approximate solution of Equation 255 for τ is[1]

$$\tau \doteq \tau(M', C_{Lo}) - 0.7\delta \quad (258)$$

which, when substituted into Equation 257, gives

$$Q \doteq \left[0.319C_{Lo} + 1.25\tau(M', C_{Lo}) - 0.125\delta\right]^2 + C_{df} \quad (259)$$

Consequently, δ should be maximized, since Q is to be minimized. (C_f is relatively small and nearly independent of τ, δ, and k, so

---

[1] Notice that $0.63\delta + \tau$ is the relationship of δ and τ in the first term on the right side of Equation 255, and that $0.70 \delta + \tau$ is the relationship in the second term. If the relationship had been $0.70 \delta + \tau$ in both terms, Equation 255 could have been solved for $0.70 \delta + \tau$ as a function of M' and C_{Lo}, which gives Equation 258. This solution is very close, because the second term is much larger than the first.

it can be ignored when Q is optimized.) Similarly, an approximate solution of Equation 255 for δ is,

$$\delta \doteq \delta(M', C_{Lo}) - 1.43\tau \qquad (260)$$

which, when substituted into Equation 257, gives

$$Q \doteq \left[0.319C_{Lo} + 0.75\,\delta(M', C_{Lo}) + 0.18\tau\right]^2 + C_{df} \qquad (261)$$

Consequently, τ should be minimized in order to minimize Q.

Summarizing, δ should be maximized and τ should be minimized. The boundary conditions on τ, δ, and k are

$$\left(\begin{array}{c}\text{Region II}\\ \sigma = 0\end{array}\right) \quad \left\{\begin{array}{l}\delta \geq 0.090k \text{ when } \tau = 0 \\[2mm] k,\ \tau,\ \delta \geq 0\end{array}\right. \qquad (262)$$

where the first condition is obtained from the previous mapping, and is the value of δ required to prevent the cavity from intersecting the lower surface of the hydrofoil; the second condition is the set of values of k, τ, and δ which are required to prevent the possibility of lower surface cavitation.

The maximum value of δ is investigated first. From Equation 153 it is seen that the maximum value of δ is $\frac{2}{\pi}C_{Lo}$ and occurs when k = 0. Setting k = 0 in Equation 254, substituting $\delta = \frac{2}{\pi}C_{Lo}$, and solving for τ where $\tau \geq 0$,

$$\tau = -0.426C_{Lo} + \sqrt{3.72M' - 0.0053C_{Lo}^2} \geq 0 \qquad (263)$$

which reduces to

$$\left(\begin{array}{c} \sigma = 0, \ k = 0 \\[1.5em] \delta = \dfrac{2}{\pi} \, C_{Lo} \end{array}\right) \ M' \geqq 0.0502 C_{Lo}^{2} \tag{264}$$

The case of minimum τ is now investigated by setting $\tau = 0$. From Equation 253 it is seen that $\delta = \dfrac{2}{\pi} (C_{Lo} - k)$ which, when substituted into Equation 254 with $\tau = 0$, gives

$$(\sigma = 0, \ \tau = 0) \quad k = 0.962 C_{Lo} \pm \sqrt{18.9 M' - 0.023 C_{Lo}^{2}} \geqq 0 \tag{265}$$

where $k \geqq 0$. The first restriction of Equation 262 that $\delta \geqq 0.090k$ must be satisfied in solving Equation 265 because $\tau = 0$. Since Equation 253 shows that $C_{Lo} = k + (\pi/2)\delta$, then $C_{Lo} \geqq 1.14k$, which is the same as $k \leqq 0.875 C_{Lo}$. When used with Equation 265, the inequality becomes

$$(\sigma = 0, \ \tau = 0) \quad 0 \leqq 0.962 C_{Lo} \pm \sqrt{18.9 M' - 0.023 C_{Lo}^{2}} = k \leqq 0.875 C_{Lo}$$
$$\tag{266}$$

Since the sign of the radical must be negative (to maximize δ in view of Equation 253), the above inequality after simplification becomes

$$(\sigma = 0, \ \tau = 0) \quad 0.0016 C_{Lo}^{2} \leqq M' \leqq 0.0502 C_{Lo}^{2} \tag{267}$$

A further condition that must be satisfied is that the radical of Equation 265 must not be imaginary; therefore,

$$M' \geqq 0.0012 C_{Lo}^{2} \tag{268}$$

which is less restrictive than Equation 267, and therefore not significant.

The results of Equations 264 and 267, together with Equations 263 and 265 show that at least two sets of equations are required to specify the design form families resulting from Subspace (d) of mission space. The corresponding regions of Subspace (d) are called Region IIc and Region IId, where

$$
\left.\begin{array}{c}
0.0502C_{Lo}^{2} \overset{\leq}{=} M' \\
\left(\begin{array}{c} \text{Region IIc} \\ \sigma = 0 \end{array}\right)
\end{array}\right\}
\begin{cases}
k = 0 \\
\delta = (2/\pi)C_{Lo} \\
\tau = 1.93\sqrt{M' - 0.0014C_{Lo}^{2}} - 0.426C_{Lo}
\end{cases}
\tag{269}
$$

$$
\left.\begin{array}{c}
0.0016C_{Lo}^{2} \overset{\leq}{=} M' \overset{\leq}{=} 0.0502C_{Lo}^{2} \\
\left(\begin{array}{c} \text{Region IId} \\ \sigma = 0 \end{array}\right)
\end{array}\right\}
\begin{cases}
\tau = 0 \\
k = 0.962C_{Lo} - 4.35\sqrt{M' - 0.0012C_{Lo}^{2}} \\
\delta = 0.024C_{Lo} + 2.76\sqrt{M' - 0.0012C_{Lo}^{2}}
\end{cases}
\tag{270}
$$

The missing region is $M' \overset{\leq}{=} 0.0016C_{L}^{2}$, which is considered now, and is called Region IIe. Note that when $M' = 0.0016C_{L}^{2}$ in Equation 270, then $k = 0.875C_{Lo}$, $\delta = 0.079C_{Lo}$, and $\tau = 0$. These are precisely the values of k, δ, and τ chosen for the previous mapping of Subspace (c) where $M' = 0$. Consequently, this hydrofoil form is selected for the entire region of $M' \overset{\leq}{=} 0.0016$, since it is the form which has the smallest possible drag coefficient and satisfies the requirements for C_{Lo} and M'. Its characteristics are

$$\begin{pmatrix} 0 \leq M' \leq 0.0016C_{Lo}^2 \\ \text{(Region IIe)} \end{pmatrix} \quad \begin{cases} \tau = 0 \\ k = 0.879C_{Lo} \\ \delta = 0.079C_{Lo} \end{cases} \qquad (271)$$

The upper and lower surfaces of the hydrofoil forms, as functions of k, δ, and τ, are

$$\begin{pmatrix} \text{Regions IIc,} \\ \text{IId, and IIe} \end{pmatrix} \begin{cases} y_u' = y_1'(x') \cdot k + y_3'(x') \cdot \delta + y_5'(x') \cdot \tau \\ y_\ell' = y_2'(x') \cdot k + y_4'(x') \cdot \delta - y_5'(x') \cdot \tau \end{cases} \qquad (272)$$

where the values of $y_1'(x')$ through $y_5'(x')$ are listed in Table 2.

Evaluation of the optimization criterion. The optimization criterion is presented in Equation 256 where C_{do} is shown as a function of k, τ, and δ. The values of k, τ, and δ are obtained from Equations 269 to 271. The frictional drag contribution to C_d is the same as that of Equation 250 for the mapping of Subspace (c) where $\sigma = 0$. Combining Equations 256 and 250

$$\begin{pmatrix} \text{Region II} \\ \sigma = 0 \end{pmatrix} \quad C_d = \left[0.319k + 1.25(\tau+\delta) \right]^2 + C_f (1 - \frac{C_{Lo}}{2})^2 \qquad (273)$$

Presentation of the mapping result. The lower graph of Figure 35 illustrates the mapping result of Subspace (d) of mission space where M' is plotted against C_L, and sketches of the corresponding design forms are superimposed. The upper graph of Figure 35 shows the values of C_{do} associated with Subspace (d).

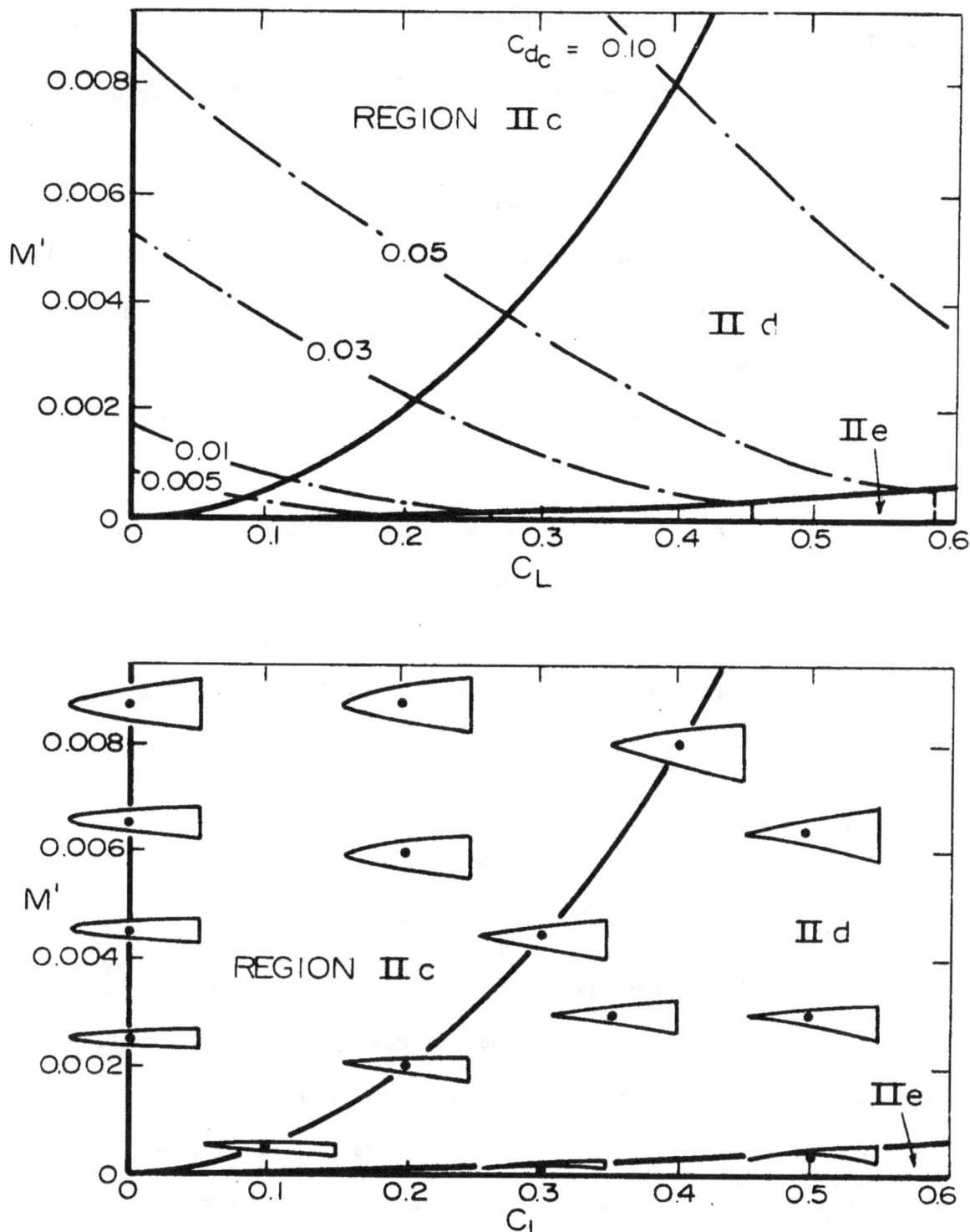

Figure 35 — Hydrofoil forms and drag coefficients mapped from Subspace (d)

Means for Increasing the Leading Edge Strength of Supercavitating Hydrofoils

One of the practical difficulties encountered in the use of supercavitating hydrofoils is leading edge vibration or failure caused by the excessively thin and sharp leading edge. This problem can be solved, in general, by exchanging the τ parabolic-type thickness for the δ wedge-type thickness which is used for Regions IId and IIe of mission space.

Considerable strength is added to the nose region when values of $\tau \gtrsim 0.10$ are used. The penalty in drag when τ is used is generally only a few percent. Note that when τ is substituted for δ, k must be increased since δ also contributes to C_{Lo} while τ does not. Keeping the values of M' and C_{Lo} fixed, the equations governing the exchange of τ for δ are seen from Equations 253 and 254 to be

$$
\text{(Region II)} \quad
\begin{cases}
\Delta\tau \doteq -\dfrac{2}{3}\,\Delta\delta \\[2em]
\Delta k = -\dfrac{\pi}{2}\,\Delta\delta
\end{cases}
\tag{274}
$$

The graphs of Reference (32) should be consulted for more accurate values of Δk and $\Delta\delta$ when τ is added, since Equation 274 provides only approximate values. The nondimensional leading edge radius, as a function of τ, is $0.5\tau^2$. Also, the nondimensional thickness produced by τ is $2\tau\sqrt{x'}$. For example, the increase in nose thickness at $x' = 0.05$ is $0.45\,\tau$, which is a considerable thickening when practical values of τ are used.

Comparison of the Lift-to-Drag Ratios of Supercavitating Hydrofoils Operating at $\sigma = 0$

As a matter of side interest, the L/D ratios of various forms of supercavitating hydrofoils are examined to determine the effect on L/D of changes in form, C_L, C_f, and M'. Sufficient thickness is added to all camber forms to make the cavity pass above the camber line; also, it is assumed that the cavity is filled with metal up to the trailing edge for calculating strength. The data for the flat plate with a flap was obtained from (26), and the rest of the data was calculated using the information from (31) and (32), and from Equations 270, 271, and 273. The results are listed in Table 3.

TABLE 3

L/D RATIOS OF VARIOUS SUPERCAVITATING HYDROFOIL FORMS AT $\sigma = 0$

Hydrofoil Form	L/D $\begin{pmatrix} M' = 0 \\ C_f = 0 \end{pmatrix}$	L/D $\begin{pmatrix} M' = 0 \\ C_f = 0 \\ C_L = 0.2 \end{pmatrix}$	L/D $\begin{pmatrix} M' = 0.0005 \\ C_f = 0 \\ C_L = 0.2 \end{pmatrix}$	L/D $\begin{pmatrix} M' = 0 \\ C_f = 0.004 \\ C_L = 0.2 \end{pmatrix}$	L/D $\begin{pmatrix} M' = 0.0005 \\ C_f = 0.004 \\ C_L = 0.2 \end{pmatrix}$
Flat plate	$1.57/C_L$	7.9	7.9	7.0	7.0
Flat plate + 25% flap	$5.21/C_L$	26.1	---a	18.4	---
Constant pressure	$6.25/C_L$	31.3	---	20.8	---
2-term camber	$7.05/C_L$	35.3	22.8	22.5	16.6
5-term camber	$7.75/C_L$	38.7	---	24.0	---

aA blank space indicates lack of data on the cross-sectional strength.

Mapping from Subspace (e) (C_L variable, M' = 0.0005, σ variable)

The selected value of M' is typical for high-speed struts and hydrofoils.

Region boundaries. As in Subspace (c), the mission space is represented by a graph of ordinate σ versus abscissa C_L and splits into an upper region, called Region I, which maps into fully-wetted hydrofoils, and a lower region, called Region II, which maps into cavitating hydrofoils.

The previous discussions show that the minimum-drag hydrofoils corresponding to Region I will be ellipses whose meanlines are cambered using the NACA a = 1.0 uniform pressure meanline. Assuming that the camber is small, the strength of a cambered ellipse is approximately the same as that of the corresponding uncambered ellipse.

The boundary between Regions I and II is determined by the incipient cavitation number of the Region I hydrofoils, as in previous mappings. According to (21), the pressure on the upper surface of a hydrofoil can be calculated if the upper surface velocity U_u is known, where

$$(\text{Region I}) \qquad U_u = U + u + u_t \qquad (275)$$

where u = circulation velocity = $\frac{1}{4}C_L U$ (Equation 228), and u_t = added velocity due to the thickness of the ellipse where $u_t = \frac{\sigma_o}{2} U$ = $U \sqrt{10.2 M'}$ (Reference 27 and Equation 199). As defined earlier, σ_o designates the value of σ at C_L = 0 for the elliptical cavity used in generating the thickness distribution of an elliptical hydrofoil.

Equations 226 and 275 show that

$$\text{(Region I)} \qquad \frac{P-P_u}{\frac{1}{2}\rho U^2} = \left(\frac{U_u}{U}\right)^2 - 1 = \left(1 + \frac{u}{U} + \frac{u_t}{U}\right)^2 - 1 \qquad (276)$$

Substituting the expressions for u and u_t into Equation 276,

$$\frac{P-P_u}{\frac{1}{2}\rho U^2} = \left(1 + \frac{C_L}{4} + \sqrt{10.2M'}\right)^2 - 1 \qquad (277)$$

If $C_L/4 \ll 1$ and $\sqrt{10.2M'} \ll 1$, then Equation 277 becomes

$$\text{(Region I)} \qquad \frac{P-P_u}{\frac{1}{2}\rho U^2} \doteq \frac{C_L}{2} + \sqrt{40.8M'} \qquad (278)$$

Since the incipient cavitation number σ_{cr} is defined as the value of Equation 278 when $P_u = P_v$, the equation of the boundary line (where $\sigma = \sigma_{cr}$) for $M' = 0.0005$, is

$$\binom{\text{Region I to II}}{\text{boundary}} \qquad \sigma \doteq \frac{C_L}{2} + \sqrt{40.8M'} \qquad (279)$$

which for $M' = 0.0005$, becomes

$$\binom{\text{Region I to II boundary}}{M' = 0.0005} \qquad \sigma \doteq \frac{C_L}{2} + 0.143 \qquad (280)$$

Mapping from Region I. The ordinates of a basic thickness distribution and the NACA $a = 1.0$ meanline can be superimposed (21), assuming the ordinates are small. Therefore, the Region I forms are

$$\text{(Region I)} \qquad \begin{cases} y_u' = y_e' + y_m' \\[2mm] y_\ell' = -y_e' + y_m' \end{cases} \qquad (281)$$

where y_e' is the equation for the elliptical semi-thickness distribution rewritten from Equation 203 as $y_e' = \sqrt{10.2M'(x'-x'^2)}$, and y_m' is the NACA a = 1.0 meanline given by Table 1 where $y_m' = y_o'(x') \cdot C_L$. Substituting these relations, Equation 281 becomes

$$(\text{Region I}) \quad \begin{cases} y_u' = \sqrt{10.2M'(x'-x'^2)} + y_o'(x') \cdot C_L \\[2ex] y_\ell' = -\sqrt{10.2M'(x'-x'^2)} + y_o'(x') \cdot C_L \end{cases} \quad (282)$$

Substituting M' = 0.0005, Equation 282 becomes the mapping relationship for Region I of Subspace (e),

$$\begin{pmatrix} \text{Region I} \\ M' = 0.0005 \end{pmatrix} \quad \begin{cases} y_u' = 0.0715\sqrt{x'-x'^2} + y_o'(x')\,C_L \\[2ex] y_\ell' = -0.0715\sqrt{x'-x'^2} + y_o'(x')\,C_L \end{cases} \quad (283)$$

The equation for the optimization criterion $Q = C_d$ for Region I is the same as in Subspace (b), since the effects of camber do not contribute to C_{df} if C_L and M' are small. From Equation 224,

$$(\text{Region i}) \quad C_d = 2C_f\left(\frac{2}{\sqrt{0.098}}\sqrt{M'} + 1\right) \quad (284)$$

Substituting M' = 0.0005,

$$\begin{pmatrix} \text{Region I} \\ M' = 0.0005 \end{pmatrix} \quad C_d = 2.29\,C_f \quad (285)$$

Boundary conditions for the mapping from Region II. The mapping from Region II must match the boundary conditions established by the previous mappings. Furthermore, it is reasonable to expect the Region II forms to merge smoothly into the Region I forms at the boundary, because they did so in all of the previous mappings.

Figure 36 was drawn to illustrate the three-dimensional mapping conditions which resulted from the previous mappings. Notice how the previous mappings form boundary conditions for the new mapping along three sides of the new region being considered. Therefore, Figure 36 illustrates another advantage of this design procedure; namely, that the more complex mapping problems are made more tractable by the boundary conditions established by simpler mappings. The form of a hydrofoil corresponding to any point in Subspace (e) can now be approximated after a brief study of Figure 36; this preliminary determination of design form would not have been possible earlier.

The equations for the forms corresponding to the σ and C_L axes in Subspace (e) can be obtained from the mappings of Subspaces (b) and (d), respectively, by substituting $M' = 0.0005$. The boundary point between Regions I and IIa on the σ axis is obtained from Equation 199 as $\sigma_o = \sqrt{40.8M'} = 6.39\sqrt{M'} = 0.143$ where σ_o designates the value of σ when $C_L = 0$. The boundary point between Regions IIa and IIb on the σ axis is obtained from Equation 218 as $\sigma_o = \sqrt{10.2M'}$ $= 3.19\sqrt{M'} = 0.0715$. The point $\sigma = 0$ is the boundary point on the σ axis separating Regions IIb and IIc. The boundary point on the C_L axis in Subspace (e) between Regions IIc and IId is obtained from Equation 269 as $C_{Lo} = 4.46\sqrt{M'}$. Finally, the boundary point between Regions IId and IIe is $C_{Lo} = 25\sqrt{M'}$ (Equation 270).

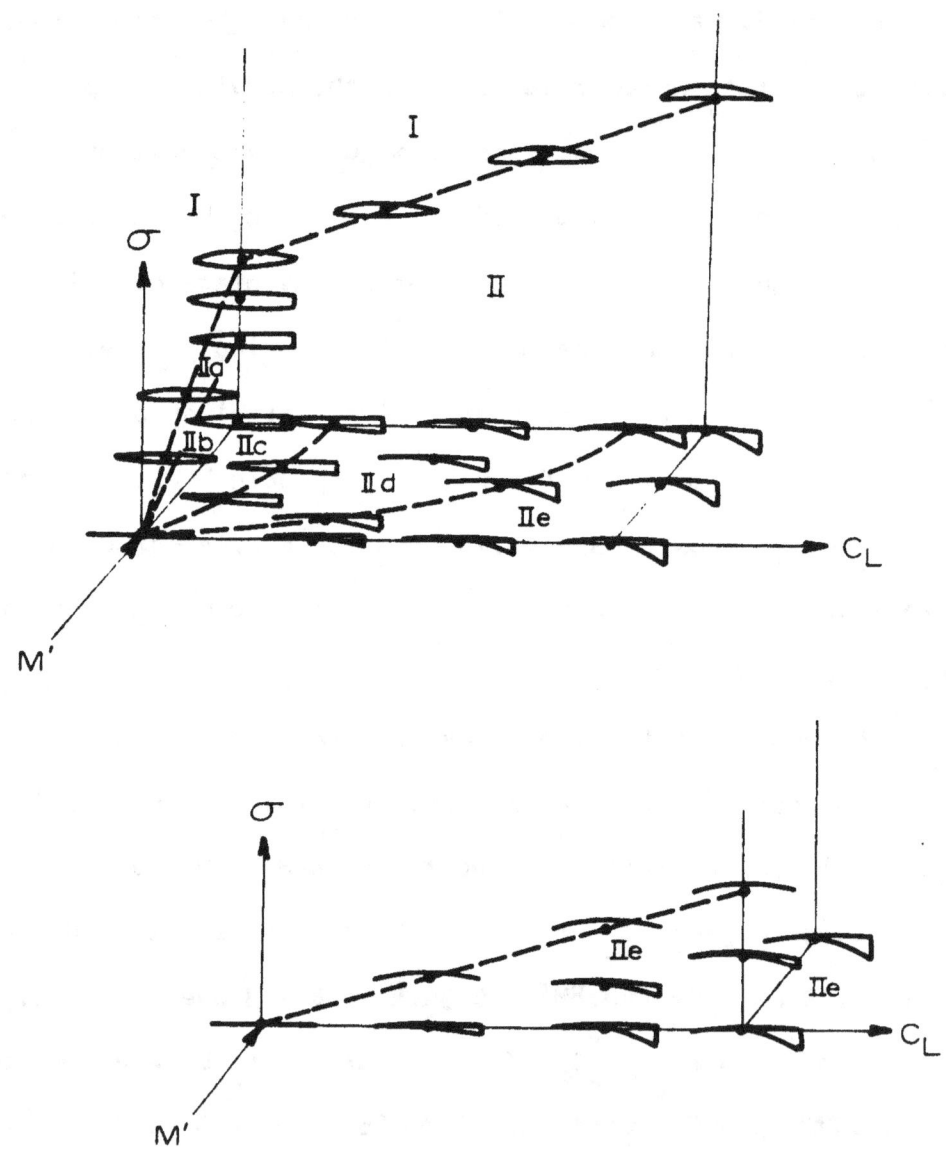

Figure 36 — Illustration of conditions imposed on the mapping from Subspace (e)

General form of designs corresponding to Region II and the
region boundaries. The mapping from points within Region II is now
considered. The literature contains no examples of forms that will
help in this mapping. However, a study of Figure 36 and of Figure
34 provides clues. Notice in Figure 34 that families of forms exist
along lines which lie parallel to the boundary line between Regions
I and II; the forms of each such family have the same thickness
distribution, but varying camber. After studying Figure 36, it is
seen that similar families of forms (having the same thickness
distribution, but varying camber) might correspond to lines paralleling the boundary line between Regions I and II in Subspace (e) where
$M' = 0.0005$. This observation is the key to the desired mapping.

Consider an extension of Regions IIa and IIb into the space
of Subspace (e). The equation of a line which parallels the
Region I — II boundary is seen from Equation 279 to be

$$\sigma = \sigma_0 + \tfrac{1}{2}C_L \tag{286}$$

where σ_0 is the intercept of any given line with the σ axis. Therefore, the hypothesized family of forms corresponding to the line of
Equation 286 would consist of forms which have the thickness distribution corresponding to the point $\sigma = \sigma_0$ superimposed on the NACA
$a = 1.0$ meanline which corresponds to C_L. Notice that any such form
would have: (a) uniform pressure on the upper surface which is
exactly equal to depth pressure, (b) exactly the specified strength,
(c) minimum-drag thickness distribution, (d) the desired lift

coefficient, and (e) no drag penalty for the lift.[1] These conditions insure that the selected form has the lowest drag of any form which can correspond to a given point in the selected region.

In view of the above discussion, the boundary between the extension of Region IIa and Region I is

$$\left(\begin{array}{l}\text{Boundary between}\\\text{Regions I and IIa}\end{array}\right) \quad \sigma = 6.39 \ \sqrt{M^T} + \tfrac{1}{2}C_L \qquad (287)$$

The boundary between the extension of Regions IIa and IIb is

$$\left(\begin{array}{l}\text{Boundary between}\\\text{Regions IIa and IIb}\end{array}\right) \quad \sigma = 3.19 \ \sqrt{M^T} + \tfrac{1}{2}C_L \qquad (288)$$

and the boundary between extended Regions IIb and IIc is

$$\left(\begin{array}{l}\text{Boundary between}\\\text{Regions IIb and IIc}\end{array}\right) \quad \sigma = \tfrac{1}{2}C_L \qquad (289)$$

The forms corresponding to Regions IIc to IIe are derived in a similar manner. The equation of a line which intersects the C_L axis at C_{Lo} and parallels the Region I to Region II boundary is

$$\sigma = \tfrac{1}{2} \ (C_L - C_{Lo}) \qquad (290)$$

where $C_L \gtrapprox C_{Lo}$. The form which corresponds to any point along this line consists of the form designed for C_{Lo} superimposed on the NACA $a = 1.0$ meanline, where the meanline lift coefficient is obtained from Equation 290 as

---

[1] The cavity drag resulting from lift produced by pure camber is zero (38).

$$\left(\begin{array}{c} \text{Regions} \\ \text{IIc to IIe} \end{array}\right) \quad \left(C_L\right)_{a=1.0} = C_L - C_{Lo} = 2\sigma \tag{291}$$

The boundary between the extension of Regions IIc and IId is derived from the C_L axis intercept of $C_{Lo} = 4.46 \sqrt{M'}$ found previously, and is

$$\left(\begin{array}{c} \text{Boundary between} \\ \text{Regions IIc and IId} \end{array}\right) \quad \sigma = -2.23 \sqrt{M'} + \tfrac{1}{2}C_L \tag{292}$$

Similarly,

$$\left(\begin{array}{c} \text{Boundary between} \\ \text{Regions IId and IIe} \end{array}\right) \quad \sigma = -12.5 \sqrt{M'} + \tfrac{1}{2}C_L \tag{293}$$

Specific forms corresponding to Region II. The specific form corresponding to any point in Region II can be obtained from the general forms which were just derived. Considering families IIa and IIb first, the values of σ_o and M' are sufficient to specify the thickness distribution which is given by Equation 217. The camber is given by Equation 231. The value of σ_o is the intercept with the σ axis of the line described earlier which is the locus of the family of cambered forms, all of which have the same thickness distribution. This intercept is obtained from Equation 286 as

$$\sigma_o = \sigma - \tfrac{1}{2}C_L \tag{294}$$

Substituting Equation 294 into Equation 217 and adding to this the camber of Equation 231, gives

(Region IIa)
$$y_u' = \frac{\left(\sigma - \frac{C_L}{2}\right)}{2} \sqrt{\frac{2\sqrt{M'}}{\frac{C_L}{\sqrt{C_1}}\left(\sigma - \frac{C_L}{2}\right)} x' - (x')^2} + y_o'(x') \cdot C_L$$

$$y_\ell' = -\frac{\left(\sigma - \frac{C_L}{2}\right)}{2} \sqrt{\frac{2\sqrt{M'}}{\frac{C_L}{\sqrt{C_1}}\left(\sigma - \frac{C_L}{2}\right)} x' - (x')^2} + y_o'(x') \cdot C_L \qquad (295)$$

and

(Region IIb)
$$y_u' = \frac{\left(\sigma - \frac{C_L}{2}\right)}{2} \sqrt{\left[\frac{M'}{C_1\left(\sigma - \frac{C_L}{2}\right)^2} + 1\right] x' - (x')^2} + y_o'(x') \cdot C_L$$

$$y_\ell' = -\frac{\left(\sigma - \frac{C_L}{2}\right)}{2} \sqrt{\left[\frac{M'}{C_1\left(\sigma - \frac{C_L}{2}\right)^2} + 1\right] x' - (x')^2} + y_o'(x') \cdot C_L \qquad (296)$$

where C_1 is obtained from Figure 32.

Similarly, the basic forms for Regions IIc to IIe are obtained from Equations 269 to 272 if C_{Lo} and M' are known. The NACA a = 1.0 meanline defined by Equation 231 is added to each basic form, where the meanline lift coefficient is given by Equation 291 as 2σ. The value of C_{Lo} is obtained from Equation 290 as

$$C_{Lo} = 2\sigma - C_L \qquad (297)$$

Substituting Equation 297 into Equations 269 to 272 and adding the meanline of Equation 231, where the meanline lift coefficient is 2σ, gives the forms for Regions IIc to IIe expressed in terms of k, τ, and δ as

$$
\text{(Region IIc)} \begin{cases}
y_u' = y_3'(x') \cdot \delta + y_5'(x') \cdot \tau + y_o'(x') \cdot 2\sigma \\[8pt]
y_\ell' = y_4'(x') \cdot \delta - y_5'(x') \cdot \tau + y_o'(x') \cdot 2\sigma \\[8pt]
k = 0 \\[8pt]
\delta = \dfrac{2}{\pi}(C_L - 2\sigma) \\
\tau = 1.93\sqrt{M' - 0.0014(C_L - 2\sigma)^2} - 0.426(C_L - 2\sigma)
\end{cases} \tag{298}
$$

$$
\text{(Region IId)} \begin{cases}
y_u' = y_1'(x') \cdot k + y_3'(x') \cdot \delta + y_o'(x') \cdot 2\sigma \\[8pt]
y_\ell' = y_2'(x') \cdot k + y_4'(x') \cdot \delta + y_o'(x') \cdot 2\sigma \\[8pt]
k = 0.962(C_L - 2\sigma) - 4.35\sqrt{M' - 0.0012(C_L - 2\sigma)^2} \\[8pt]
\delta = 0.024(C_L - 2\sigma) + 2.76\sqrt{M' - 0.0012(C_L - 2\sigma)^2} \\[8pt]
\tau = 0
\end{cases} \tag{299}
$$

$$
\text{(Region IIe)} \begin{cases}
y_u' = y_1'(x') \cdot k + y_3'(x') \cdot \delta + y_o'(x') \cdot 2\sigma \\[8pt]
y_\ell' = y_2'(x') \cdot k + y_4'(x') \cdot \delta + y_o'(x') \cdot 2\sigma \\[8pt]
k = 0.875(C_L - 2\sigma) \\[8pt]
\delta = 0.079(C_L - 2\sigma) \\[8pt]
\tau = 0
\end{cases} \tag{300}
$$

The specific forms for Subspace (e) are obtained by substituting $M' = 0.0005$ into Equations 295, 296, and 298 to 300.

Drag coefficients of the Region II forms. The cavity drag of the Region IIa forms is derived in Appendix E as

$$
\text{(Region IIa)} \qquad C_{dc} = \frac{\pi}{4}\sigma\frac{t}{c} \tag{301}
$$

where t/c is given in Figure 42 of Appendix E as a function of

$\sigma_o^2/M' = (\sigma - C_L/2)^2/M'$. Since σ, C_L and M' are known in a typical hydrofoil cross-section problem, C_{dc} can be readily calculated. The total drag coefficient for the Region II forms is seen by Equations 207 and 250 to be

$$\text{(Region II)} \quad C_d = C_{dc} + C_{df} = C_{dc} + C_f\left(1 - \frac{C_L}{2} + \frac{\sigma}{2}\right) \quad (302)$$

The cavity drag coefficient for the forms corresponding to Regions IIb to IIe is shown by Equation 242 to be

$$C_{dc} = C_{do} + \frac{\frac{\pi}{4}\left(\sigma \frac{t}{c}\right)^2}{\sigma \frac{t}{c} + 1.5\, C_{do}} \quad (303)$$

The values of C_{do} and t/c for the Region IIb forms can be obtained as a function of σ_o^2/M' from Figure 42 in Appendix E. The value of C_{do} for the Region IIc to IIe forms is given by Equation 256 as

$$\text{(Regions IIc to IIe)} \quad C_{do} = \left[0.319k + 1.25(\tau+\delta)\right]^2 \quad (304)$$

where k, τ, and δ are obtained from Equations 298 to 300. The values of t/c for the Region IIc to IIe forms are obtained from Table 2 where

$$\text{(Regions IIc to IIe)} \quad \frac{t}{c} = 1.92k + 1.68\delta + 2.00\tau \quad (305)$$

Presentation of the mapping. The lower graph of Figure 37 shows Regions I to IIe and some corresponding forms plotted as a function of σ and C_L where $M' = 0.0005$. The upper graph of Figure 37 shows the value of C_{dc} plotted in Region II and the value of

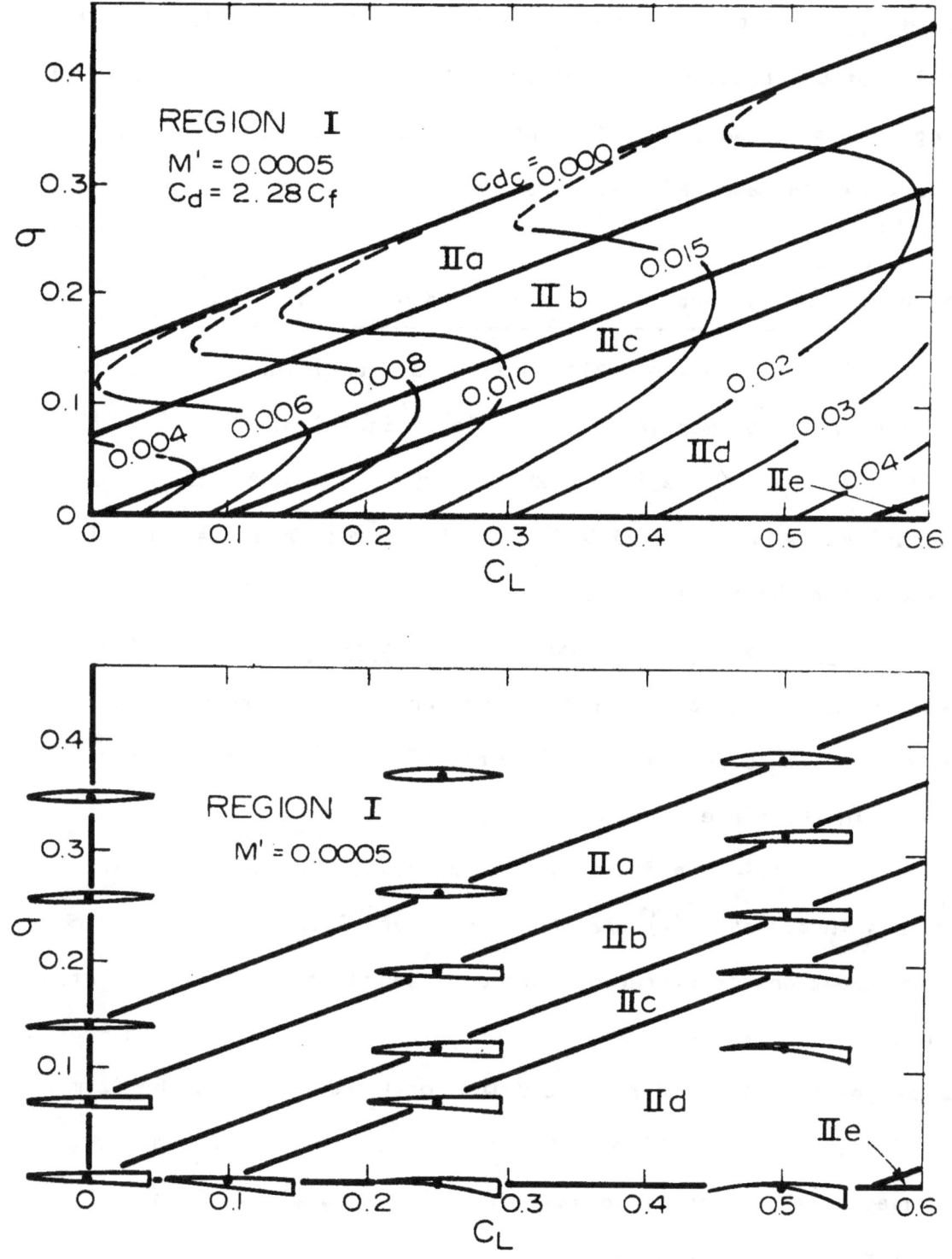

Figure 37 — Hydrofoil forms and drag coefficients mapped from Subspace (e)

$C_d = C_{df}$ plotted in Region I, where $M' = 0.0005$. Notice that the value of C_{dc} reduces as σ increases from zero, until the Region IIb, Region IIc boundary is met. The forms corresponding to this boundary are cambered parabolas.

Mapping from Subspace (f) (C_L, M', and σ are variable)

The mapping from this three-dimensional subspace was completed in the course of mapping Subspace (e). Equations 282, 284, 287 to 289, 292, 293, 295, 296, and 298 to 305, and Figures 32 and 42, describe the various boundaries, design forms, and drag coefficients needed for the complete mapping.

Illustration of the mapping boundaries and forms. The three-dimensional boundaries in Subspace (f) of mission space are shown in Figure 38. Notice the stratification of the boundary lines in the plane where M' = constant.

The various hydrofoil forms corresponding to different points in Subspace (f) are shown in Figure 39. According to the form equations, a different form corresponds to each of the infinite number of points in Region IIa to Region IId. The forms in Region I are different for different values of C_L and M', and the forms in Region IIe are different for different values of C_L and σ. Consequently, there is no best over-all hydrofoil form, but rather an infinite variety of best forms in which the best single form depends upon the specific operating situation.

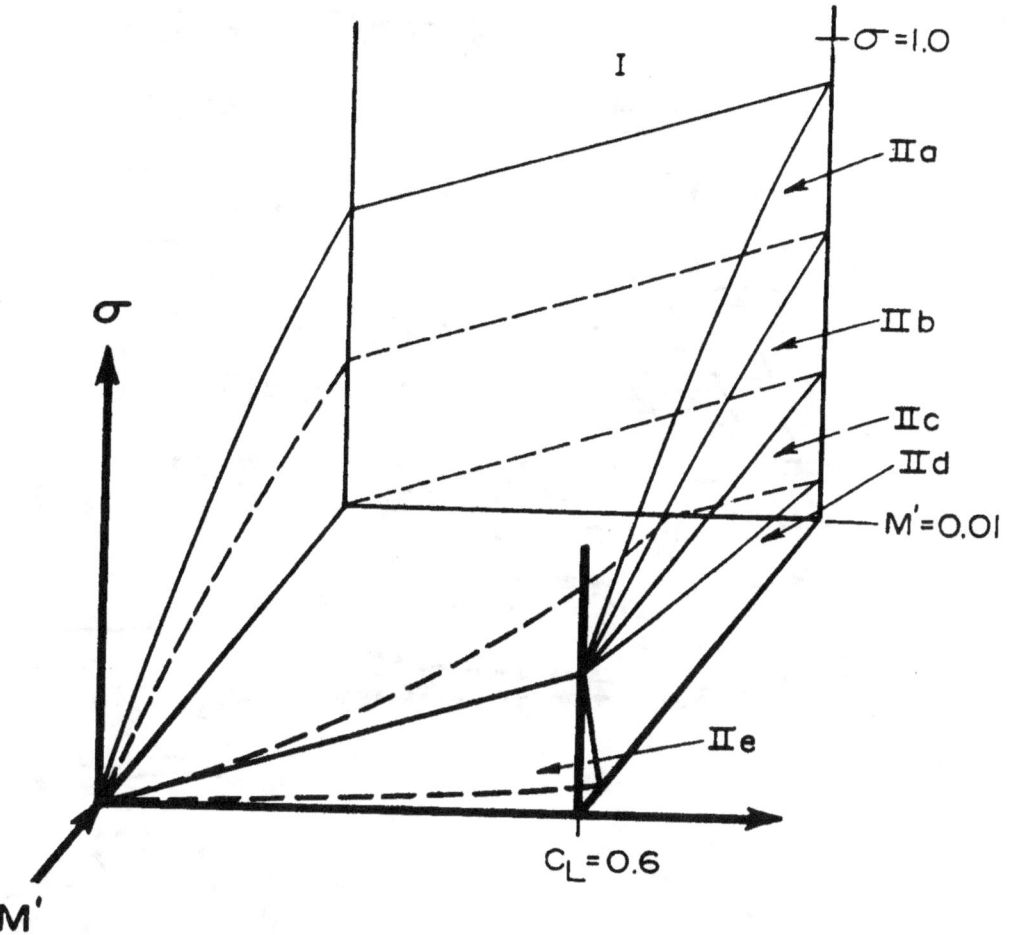

Figure 38 - Boundaries of Regions I through IIe in
three-dimensional space

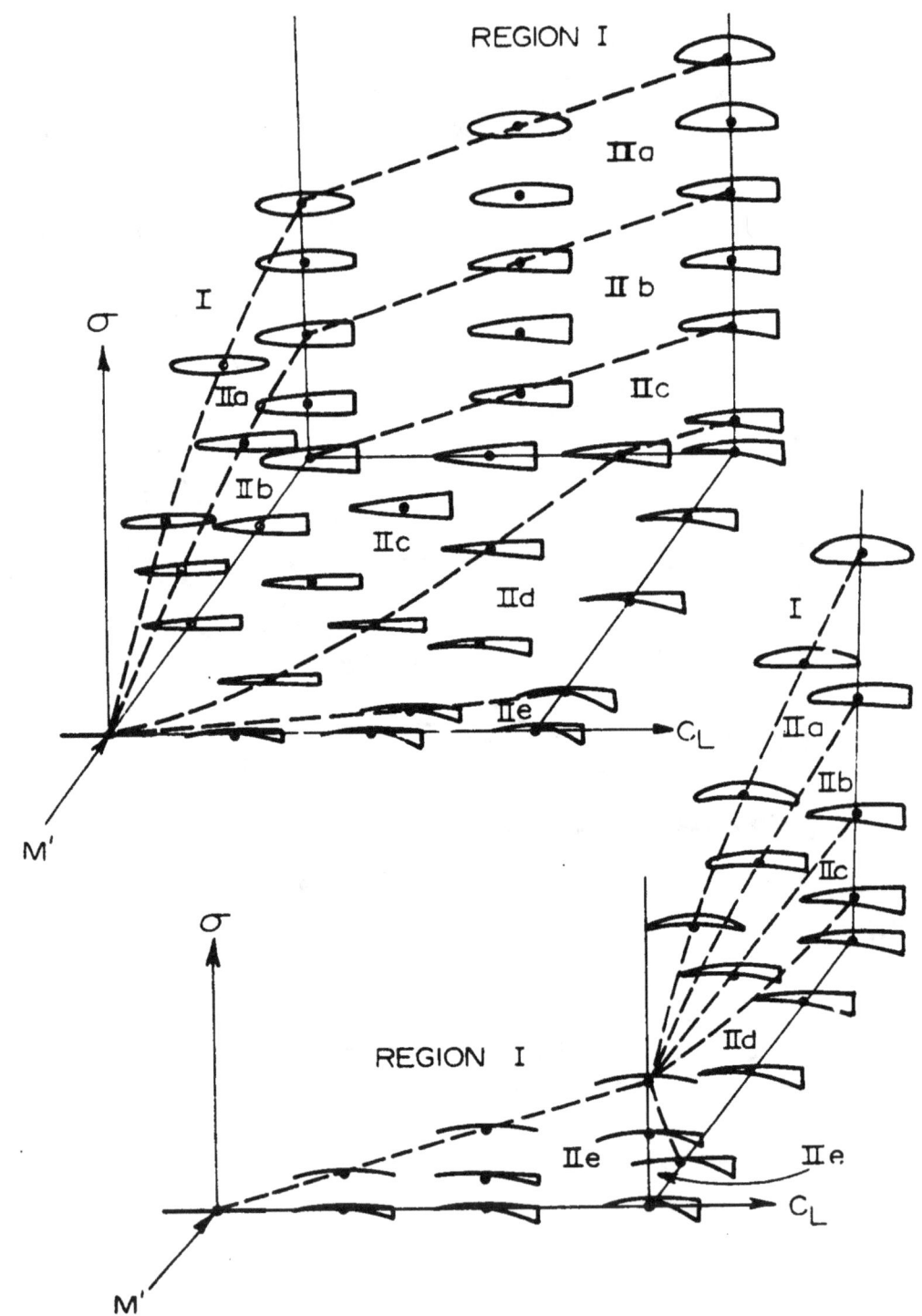

Figure 39 — Illustration of a three-dimensional mapping into hydrofoil forms

Notice that a straight or curved line anywhere in Figure 39 is the locus of a smoothly-varying set of design forms. One can easily find the set of cambered parabolas, ellipses, semi-ellipses, flat-bottomed forms, cambered lines, etc. Figure 40 shows the various families of hydrofoil forms and how they relate to each other.

Transformation of the Three-Dimensional Subspace of Mission Space into a One-Dimensional Subspace

A most remarkable result now becomes evident which was not apparent earlier. Notice that the region boundaries given by Equations 287 to 289 and 292 to 293 can be expressed as a function of only one parameter which is denoted by K where

$$K = \frac{\sigma - C_L/2}{\sqrt{M'}} = \frac{\sigma_0}{\sqrt{M'}} \tag{306}$$

The region boundaries are listed as a function of K in Table 4.

The simplification introduced for the description of the boundaries can be extended to the description of the hydrofoil forms and the drag coefficients. The following definitions are introduced:

$$\left.\begin{array}{l} \overline{C}_L = C_L / \sqrt{M'} \\[4pt] \overline{\sigma} = \sigma / \sqrt{M'} \\[4pt] \overline{y'} = y' / \sqrt{M'} = (y/c) / \sqrt{M'} \\[4pt] \overline{t'} = t' / \sqrt{M'} = (t/c) / \sqrt{M'} \\[4pt] \overline{\delta} = \delta / \sqrt{M'} \\[4pt] \overline{\tau} = \tau / \sqrt{M'} \\[4pt] \overline{k} = k / \sqrt{M'} \\[4pt] \overline{C}_{dc} = C_{dc} / M' \\[4pt] \overline{C}_{do} = C_{do} / M' \end{array}\right\} \tag{307}$$

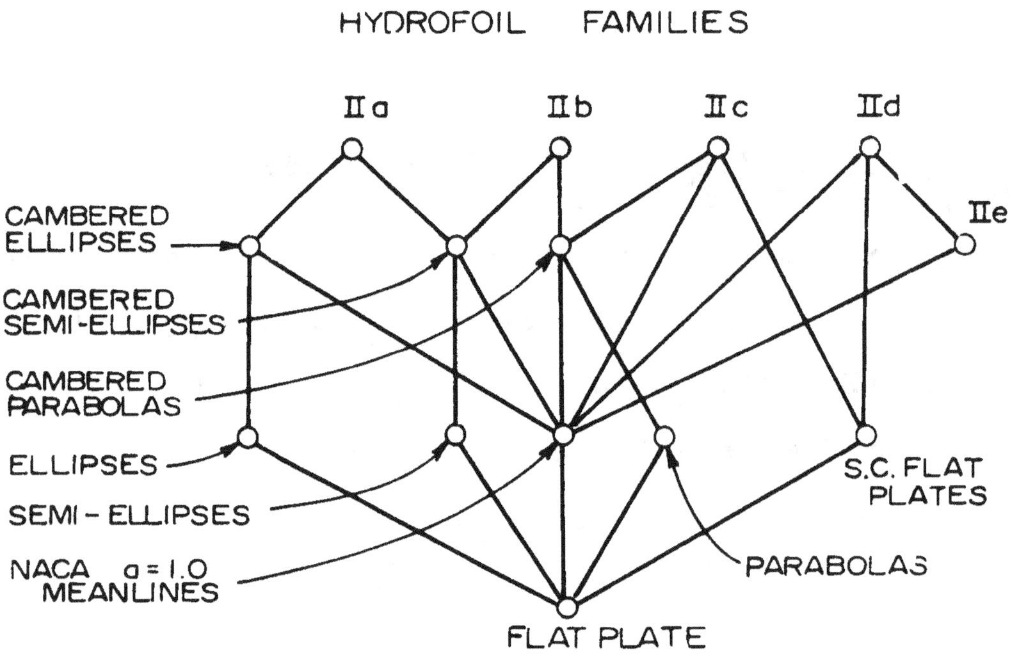

Figure 40 - The relationship of the different hydrofoil families corresponding to Subspace (f)

TABLE 4

REGION BOUNDARIES FOR SUBSPACE (f) AS A FUNCTION OF K

Region Boundary	Equation
I to IIa	$K = 6.39$
IIa to IIb	$K = 3.19$
IIb to IIc	$K = 0$
IIc to IId	$K = -2.23$
IId to IIe	$K = -12.5$

The expressions for the hydrofoil forms given by Equation 282, 295, 296, and 298 to 300 are transformed into the new parameters and listed in Table 5. The values for C_1 are shown in Figure 32 as a function of K^2. The expressions for C_{do} and C_{df} are listed in Table 6, and were obtained from Equations 301, 302, 304, 305, and Table 5.

TABLE 6

HYDROFOIL DRAG COEFFICIENTS FOR THE FORMS
CORRESPONDING TO SUBSPACE (f)

Region	$\overline{C}_{do}$ (or $\overline{C}_{dc}$)	C_{df}
I	0	$2C_f(1+6.39\sqrt{M'})$
IIa	$\left(\overline{C}_{dc} = \frac{\pi}{4}\frac{\sigma}{\sqrt{C_1 M'}}\right)$	$C_f\left(1 - \frac{C_L}{2} + \frac{\sigma}{2}\right)^2$
IIb	See Figure 42	"
IIc	$\left[2.41\sqrt{1-0.0056K^2}-0.525K\right]^2$	"
IId	$\left[2.06\sqrt{1-0.0048K^2}-0.674K\right]^2$	"
IIe	$0.572K^2$	"

TABLE 5

HYDROFOIL FORM CHARACTERISTICS CORRESPONDING TO SUBSPACE (f)

Region	Form Equation	$\bar{t}' = \dfrac{t/c}{\sqrt{M'}}$	$\dfrac{t_b}{t}$
I	$\bar{y}_u' = 3.19\sqrt{x' - (x')^2} + y_o'\overline{C_L}$ $\bar{y}_\ell' = -3.19\sqrt{x' - (x')^2} + y_o'\overline{C_L}$	3.19	0
IIa	$\bar{y}_u' = \dfrac{K}{2}\sqrt{\dfrac{2x'}{K\sqrt{C_1}} - (x')^2} + y_o'\overline{C_L}$ $\bar{y}_\ell' = -\dfrac{K}{2}\sqrt{\dfrac{2x'}{K\sqrt{C_1}} - (x')^2} + y_o'\overline{C_L}$	$\dfrac{1}{\sqrt{C_1}}$	$K\sqrt{C_1}\sqrt{\dfrac{2}{K\sqrt{C_1}} - 1}$
IIb	$\bar{y}_u' = \dfrac{K}{2}\sqrt{\left(\dfrac{1}{K^2C_1} + 1\right)x' - (x')^2} + y_o'\overline{C_L}$ $\bar{y}_\ell' = -\dfrac{K}{2}\sqrt{\left(\dfrac{1}{K^2C_1} + 1\right)x' - (x')^2} + y_o'\overline{C_L}$	$\dfrac{1}{\sqrt{C_1}}$	1.0
IIc	$\bar{y}_u' = y_3' \bar{\delta} + y_5' \bar{\tau} + 2y_o' \bar{\sigma}$ $\bar{y}_\ell' = y_4' \bar{\delta} - y_5' \bar{\tau} + 2y_o' \bar{\sigma}$ $\bar{\delta} = -1.272K$ $\bar{\tau} = 1.93\sqrt{1 - 0.0056K^2} + 0.85K$ $\bar{k} = 0$	$-0.436K$ $+3.86\sqrt{1 - 0.0056K^2}$	1.0
IId	$\bar{y}_u' = y_1' \bar{k} + y_3' \bar{\delta} + 2y_o' \bar{\sigma}$ $\bar{y}_\ell' = y_2' \bar{k} + y_4' \bar{\delta} + 2y_o' \bar{\sigma}$ $\bar{k} = -1.924K - 4.35\sqrt{1 - 0.0048K^2}$ $\bar{\delta} = -0.048K + 2.76\sqrt{1 - 0.0048K^2}$ $\bar{\tau} = 0$	$-0.451K$ $+3.80\sqrt{1 - 0.0048K^2}$	1.0
IIe	$\bar{y}_u' = y_1' \bar{k} + y_3' \bar{\delta} + 2y_o' \bar{\sigma}$ $\bar{y}_\ell' = y_2' \bar{k} + y_4' \bar{\delta} + 2y_o' \bar{\sigma}$ $\bar{k} = -1.750K$ $\bar{\delta} = -0.158K$ $\bar{\tau} = 0$	$-0.602K$	1.0

The basic form characteristics which consist of $\overline{t'} = (t/c)/\sqrt{M'}$, t_b/t, $\overline{\tau}$, $\overline{k}/10$, and $\overline{\delta}$ are plotted in Figure 41 as a function of K. Also shown in Figure 41 are typical hydrofoil shapes superimposed along vertical lines which represent the region boundaries. Notice how clearly and precisely Figure 41 represents all of the hydrofoil forms and how the three-dimensional illustration in Figure 39 has been condensed into a single one-dimensional graph where the only parameter is $K = (\sigma - C_L/2)/\sqrt{M'}$. This parameter K classifies all cavitating hydrofoils and the simpler fully-wetted hydrofoils much like the specific speed parameter classifies turbomachinery. The nature of the parameter K is somewhat broader than the specific speed parameter, however, because it includes the effect of cavitation and structural strength on design form which the latter does not include.

Figure 41 can be utilized with Tables 5 and 6, Equations 302, 303, and 307, and Figures 32 and 42, to completely specify the lowest-drag hydrofoil cross sections corresponding to Subspace (e) as a function of C_L, M', and σ.

General Comments on the Design of Hydrofoil Cross Sections

The results of this hydrofoil design problem are applicable to a wide variety of operating situations. The restrictions that $R_e \gg 10^7$ and r' = 0 are not necessary if the boundary layer is turbulent; an expression for C_{df} has been included to correct all drag coefficients for R_e and r'. Not even the boundary layer state restriction is needed for the case of the Region II forms.

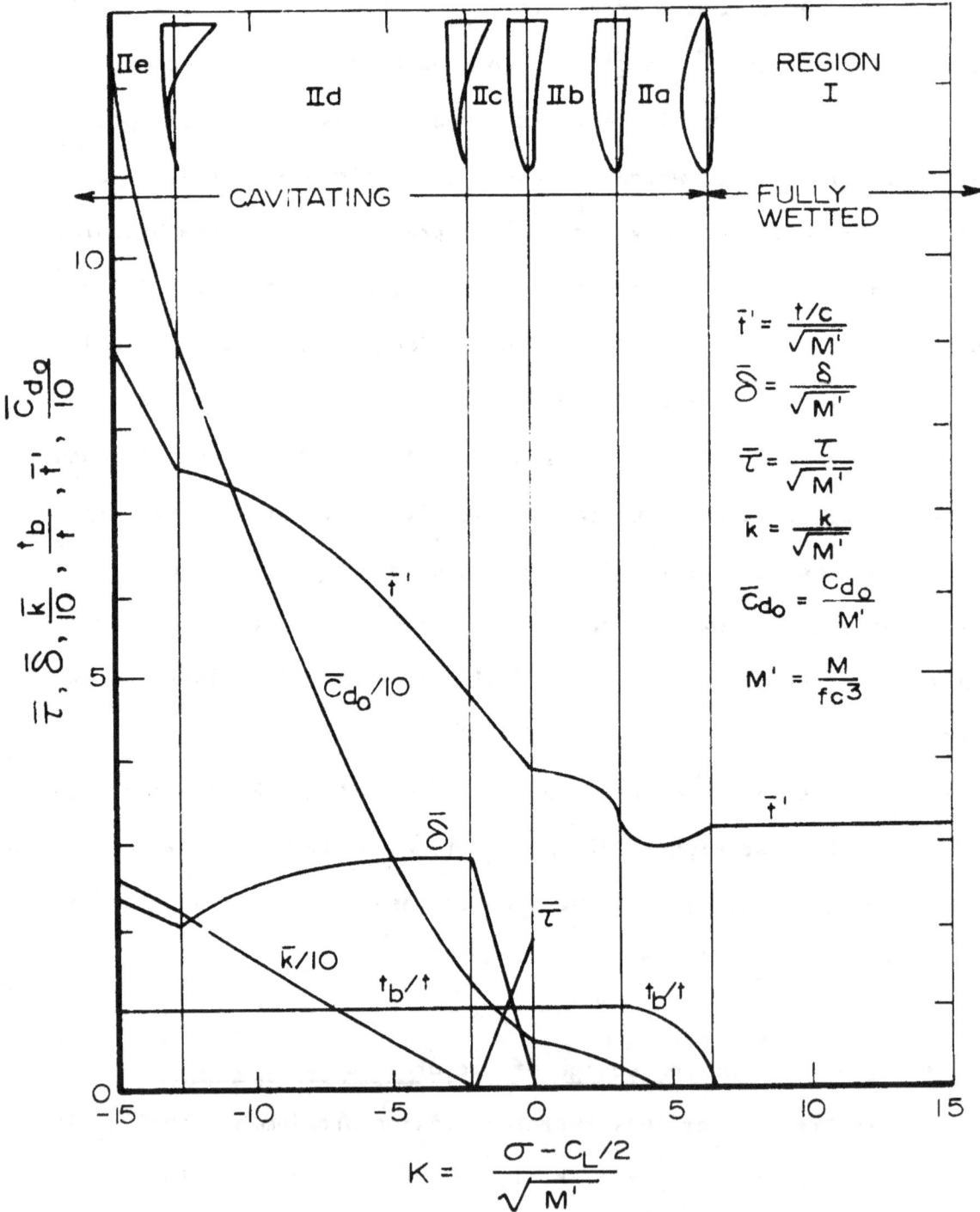

$$\bar{t}' = \frac{t/c}{\sqrt{M'}}$$

$$\bar{\delta} = \frac{\delta}{\sqrt{M'}}$$

$$\bar{\tau} = \frac{\tau}{\sqrt{M'}}$$

$$\bar{k} = \frac{k}{\sqrt{M'}}$$

$$\bar{C}_{d_o} = \frac{C_{d_o}}{M'}$$

$$M' = \frac{M}{fc^3}$$

$$K = \frac{\sigma - C_L/2}{\sqrt{M'}}$$

Figure 41 — One-dimensional representation of hydrofoil design form characteristics

The mission criterion concerning the effect of the water surface on performance is not significant, in general, since very few hydrofoils are designed to operate steadily within about two chordlengths of the surface where depth affects exist. The Froude number can be disregarded in most cases of hydrofoil cross section design because it is generally sufficiently high for cavitating hydrofoils that it has negligible effect on the cavity shape and design form. The restriction that $\Delta\alpha = 0^{\circ}$ can be relaxed to $\Delta\alpha = \pm 3^{\circ}$ or more, in general, for fully-wetted hydrofoils when the boundary layer is turbulent, without seriously influencing performance or design form unless cavitation is very critical. The effect of short periods of positive values of $\Delta\alpha$ on supercavitating hydrofoil performance or design is small; however, if $\Delta\alpha$ is to be negative, the upper surface should be undercut so that the cavity clears it at negative angles of attack. The restriction of solid sections is not serious because the designer can easily modify the specified M' to account for any amount of hollowness by using a fictiously high value of M'. Similarly, the assumption that the separation drag of the fully-wetted hydrofoils is negligible can be complied with by adding a cusp-shaped or wedge-shaped trailing edge to reduce separation of the turbulent boundary layer.[1] A final comment is that the results

of this analysis can also be made to apply to a relatively new kind

of hydrofoil form introduced by Hydronautics, Incorporated, called

a supercavitating hydrofoil with an annex (33). This form is

essentially a typical Region II hydrofoil form with an unwetted

annex extending rearward into the cavity from the trailing edge to

increase the bending strength without changing any of the performance

characteristics. Such a form can be treated in this analysis by

artificially reducing the required value of M' by perhaps thirty

percent or whatever value the designer finds reasonable in view

of the anticipated form of the hydrofoil and cavity. When the

design of the Region II form has been completed, the designer can

add the annex and check his earlier estimate of approximate annex

size and strength change.

Summarizing, the selected operating conditions for this

analysis are found to be considerably more general than they first

appeared to be.

---

[1] The better low-drag hydrofoil forms are very close in shape
to an ellipse with either a cusp-shaped or a wedge-shaped
trailing edge. For example, see the NACA 16-series and 65-
series airfoils of Reference (21). Also, a sharp trailing
edge is desirable in order to satisfy the Kutta condition
for the lifting hydrofoils. Notice that the value of M'
reduces when such a trailing edge is added; this reduction
in M' can be easily accounted for by reducing the specified
value of chordlength about 20%, or whatever value appears
reasonable for the thickness-to-chord ratio which results.
Notice that the specified value of C_L has to be changed
accordingly. This trailing edge addition only affects the
Region I forms.

Notice that the hydrofoil forms split into six different families in which each family is described by a different set of equations. Although some of the families and their boundaries in mission space are uniquely determined, the determination of others is arbitrary and depends upon the variables used in describing the hydrofoil form. For example, the boundary between Regions I and II is uniquely determined because it results from a change in physical flow condition which is not man made. On the other hand, the boundaries between Regions IId, IIe, and IIf are not unique because instead of using the variables k, τ, and δ, to represent the amount of two-term camber, parabolic thickness distribution, and δ-thickness distribution, other variables could have been used to represent other kinds of basic camber and thickness distributions. Approximately the same hydrofoil form would be found to correspond with each point in mission space, but the equations describing the forms would be different. Slight form changes and small improvements in performance will probably be found for Regions IIc, IId, and IIe as a result of future research. No changes are anticipated in the forms or boundary description corresponding to Regions I, IIa, and IIb, within the framework of the stated assumptions. Also, the classification parameter K which resulted from this analysis should remain unique.

APPENDIX C

GROUP THEORY AND DESIGN FORM TRANSFORMATIONS

The objective of this appendix is to illustrate that transformations from one design form to another within the same family
can be looked upon as elements of a group. A relatively simple
example is presented first, followed by the generalized treatment.

Transformation of Hydrofoil Cross Sections

Assume, for this example, that the mission parameters of a
generalized hydrofoil design mission consist of only C_L and M',
where C_L is the lift coefficient and M' is the nondimensional
applied bending moment. Furthermore, assume that the design objective is to design hydrofoil cross sections where the only design
parameters are the camber line $y(x)$ (i.e., center line location of
a hydrofoil cross section) and the thickness distribution $t(x)$.
Notice that the camber and thickness are both expressed as functions
of x, where x is the distance from the nose of a hydrofoil. The
pair $[y(x), t(x)]$ completely describes a hydrofoil cross section,
and the pair $[C_L, M']$ completely describes a design mission.

Assume that the physical relationships between the mission
parameters and the design parameters are the following:

$$y(x) = C_L \, y_o(x) \tag{308}$$

$$t(x) = \sqrt{M'}\, t_o(x) \tag{309}$$

where $y_o(x)$ is the (known) camber line corresponding to $C_L = 1.0$, and $t_o(x)$ is the (known) thickness distribution corresponding to $M' = 1.0$.

Let p designate a given, but arbitrary, design mission. Using Equations 308 and 309 , the corresponding design form is designated by $[y_p(x),\ t_p(x)]$, where

$$y_p(x) = C_{L_p}\, y_o(x) \tag{310}$$

$$t_p(x) = \sqrt{M'_p}\, t_o(x) \tag{311}$$

If C_L and M' are looked upon as coordinates in a two-dimensional Euclidean space called mission space, any pair of real numbers corresponding to $[C_{L_p},\ M'_p]$ represents a point in the mission space. Similarly, any pair of real functions corresponding to $[y_p(x),\ t_p(x)]$ represents a point in a two-dimensional Euclidean space called design space. Equations 310 and 311 therefore represent a mapping from an arbitrary point in mission space to the corresponding point in design space.

The mapping from M'_p to $t_p(x)$ given by Equation 311 is linearized by transforming the mission space into a new mission space where the coordinates are C_L and $\sqrt{M'}$ rather than C_L and M'. Although Equation 311 remains unchanged, it may now be considered to be a linear mapping from $\sqrt{M'_p}$ to $t_p(x)$, where a point in the new mission space is represented by $[C_{L_p},\ \sqrt{M'_p}]$.

Let $p = 1$ designate a given, but arbitrary, point in the new mission space. Let an arbitrary second point in the new mission space be designated by $p = 2$ where the second point is determined by the change in coordinates ΔC_{L_q} and $\Delta (\sqrt{M'})_q$ where

$$C_{L_2} = C_{L_1} + \Delta C_{L_q} \tag{312}$$

$$\sqrt{M'_2} = \sqrt{M'_1} + \Delta (\sqrt{M'})_q \tag{313}$$

To simplify the nomenclature, let

$$\Delta C_{L_q} = r_q \tag{314}$$

$$\Delta (\sqrt{M'})_q = s_q \tag{315}$$

where r_q and s_q are real numbers, and q designates a given, but arbitrary, change in mission space coordinates. Substituting Equations 314 and 315, Equations 312 and 313 become

$$C_{L_2} = C_{L_1} + r_q \tag{316}$$

$$\sqrt{M'_2} = \sqrt{M'_1} + s_q \tag{317}$$

Equations 310 to 317 show that the relationship between the two corresponding design forms is

$$y_2(x) = y_1(x) + r_q \, y_0(x) \tag{318}$$

$$t_2(x) = t_1(x) + s_q \, t_0(x) \tag{319}$$

where $y_1(x) = C_{L_1} y_0(x)$ and $t_1(x) = \sqrt{M_1'}\, t_0(x)$.

Let the arbitrary transformation g_q from design form one to design form two be defined by

$$[y_1(x),\ t_1(x)] \xrightarrow{g_q} [y_1(x) + r_q\, y_0(x),\ t_1(x) + s_q t_0(x)] = [y_2(x),\ t_2(x)]$$

(320)

Let two values of q be α and β, and let the binary operation $g_\alpha \circ g_\beta$ be defined as

$$[y_1(x),\ t_1(x)] \xrightarrow{g_\alpha \circ g_\beta} [y_1(x) + (r_\alpha + r_\beta)\, y_0(x),\ t_1(x) + (s_\alpha + s_\beta)\, t_0(x)]$$

(321)

Let the set of all r_q and the set of all s_q belong to different groups of real numbers under addition. The binary operation designated by o is then seen by Equations 320 and 321 to be a transformation by composition because g_β operates on the design form which results from the transformation g_α to produce a third design form, all of which belong to the same design form family. In other words,

$$g_\alpha \circ g_\beta = g_\gamma \qquad (322)$$

where $r_\alpha + r_\beta = r_\gamma$ and $s_\alpha + s_\beta = s_\gamma$.

Let e be the identity transformation of the set of all g_q whose corresponding values of r_q and s_q are zero. Equation 321 then shows that

$$g_q \circ e = e \circ g_q = g_q \qquad (323)$$

Let the inverse of g_q be defined as

$$[y_1(x), t_1(x)] \xrightarrow{g_q^{-1}} [y_1(x) - r_q y_0(x), t_1(x) - s_q t_0(x)] \quad (324)$$

It then follows from Equations 321 and 324 that

$$g_q \circ g_q^{-1} = g_q^{-1} \circ g_q = e \quad (325)$$

Also, from Equation 321 and the definitions of r_q and s_q, it follows that the operation is associative since

$$g_\alpha \circ (g_\beta \circ g_\gamma) = (g_\alpha \circ g_\beta) \circ g_\gamma \quad (326)$$

The set of g_q is therefore seen to satisfy all of the requirements for a group in view of Equations 320 to 326. Consequently, the set of all transformations from one hydrofoil form to another of a given hydrofoil family is a group.

General Design Form Transformations

Let mission space be represented by the set $\{m_i\}$ where each m_i is an independent mission parameter, and let design space be represented by the set $\{d_j\}$ where each d_j is an independent design parameter, where i and j are integers, and each set is finite. In view of Chapters II and IV, it is possible to develop the following set of functions f_j which range over all values of j:

$$d_j = f_j(\{m_i\}) \quad (327)$$

For each d_j, it is seen from Equation 327 that

$$d_{j2} - d_{j1} = f_{j2} - f_{j1} \quad (328)$$

where d_{j1} and d_{j2} are the values of the design parameter d_j for two different design forms. Rewriting,

$$d_{j2} = d_{j1} + (f_{j2} - f_{j1}) \tag{329}$$

Letting

$$f_{j2} - f_{j1} = r_{jk} \tag{330}$$

Equation 329 becomes

$$d_{j2} = d_{j1} + r_{jk} \tag{331}$$

Let the design form transformation g_{jk} be defined as

$$d_{j1} \xrightarrow{\ g_{jk}\ } d_{j1} + r_{jk} = d_{j2} \tag{332}$$

Define the identity transformation e as the transformation g_{jk} where $r_{jk} = 0$, and define the inverse transformation of g_{jk} as

$$d_{j1} \xrightarrow{\ g_{jk}^{-1}\ } d_{j1} - r_{jk} \tag{333}$$

Let the binary operation designated by o be defined as

$$d_{j1} \xrightarrow{\ g_{j1} \circ g_{j2}\ } d_{j1} + (r_{j1} + r_{j2}) \tag{334}$$

Let each set of r_{jk}, where j is fixed, belong to a separate group of real numbers under addition. From the definition of e and Equations 332 to 334, $\{g_{jk}\}$ is seen to be a group by composition because the following hold true for arbitrary elements of $\{g_{jk}\}$:

$$e \circ g_{j1} = g_{j1} \circ e = g_{j1} \qquad g_{j1} \circ g_{j2} = g_{j3}$$

$$\tag{335}$$

$$g_{j1} \circ g_{j1}^{-1} = g_{j1}^{-1} \circ g_{j1} = e \qquad g_{j1} \circ (g_{j2} \circ g_{j3}) = (g_{j1} \circ g_{j2}) g_{j3}$$

Since Equation 329 holds for each design form parameter belonging to the set $\{d_j\}$, the set of all transformations from one design form to another can be regarded as a group.

APPENDIX D

THE EFFECT OF SWEEPBACK ON THE INCIPIENT CAVITATION NUMBER OF HYDROFOILS

Consider two solid hydrofoils, one unswept and one swept-back. Let each have the same area bc, planform taper ratio τ, span b, lift L, and cross sectional shape. Let the thickness-to-chord ratio be constant everywhere. Each hydrofoil will then have the same overall lift coefficient C_L, aspect ratio A_r, mean chordlength c in the free-stream direction, and essentially the same induced drag.

The maximum bending stress at the root section of an optimized unswept hydrofoil is given by Equation 146 (without the inequality sign, and assuming $W_s = W_x = 0$) as

$$f = \frac{C_4 bL}{4C_1 C_3^3 t^2 c} \qquad (336)$$

where $c_4 b/2$ is the distance from the root to the semispan center of pressure, C_1 is the nondimensional section modulus coeff'cient, t is the mean thickness, and $C_3 = t_0/t = c_0/c$ where the subscript o refers to the root section. The bending stress for the sweptback foil is set equal to the unswept value to compare thickness, and is

$$f = \frac{C_4 (b/\cos\lambda)L}{4C_1 C_3^3 t_\lambda^2 \ c \ \cos\lambda} \qquad (337)$$

where λ is the sweepback angle and t_λ is the mean thickness of the sweptback foil. Equating the bending stress in Equations 336 and 337, it is seen that the sweptback foil must have a greater mean thickness since

$$t_\lambda = t/\cos\lambda \qquad (338)$$

Assuming that the cross section is one of the NACA 16-series airfoils, the incipient cavitation number of the unswept foil is given by Equation 152 as

$$\sigma_{cr} = 2.45 \, t/c + 0.56 \, C_L \qquad (339)$$

(The specific cross sectional shape can be arbitrary in this proof, but the NACA 16-series airfoil is chosen because of its excellent cavitation resistance.) The incipient cavitation number of the sweptback hydrofoil is shown by (21) to be

$$(\text{swept}) \; \sigma_{cr} = \frac{\frac{1}{2}\rho(U\cos\lambda)^2}{\frac{1}{2}\rho U^2} \cdot \sigma_{\lambda cr} = \sigma_{\lambda cr}\cos^2\lambda \qquad (340)$$

where $\sigma_{\lambda cr}$ is the cavitation number based upon the spanwise cross section and the component of U which is perpendicular to the swept span. Using Equation 339, $\sigma_{\lambda cr}$ is

$$\sigma_{\lambda cr} = 2.45 \, t_\lambda/c \, \cos\lambda + 0.56 \, C_{L\lambda} \qquad (341)$$

where $C_{L\lambda}$ is the design lift coefficient of the spanwise cross section, and is related to C_L by equating the lift of the two hydrofoils as follows:

$$L = C_L \ bc \ \tfrac{1}{2}\rho U^2 = C_{L\lambda} \ bc \ \tfrac{1}{2}\rho (U\cos\lambda)^2 \qquad (342)$$

Solving,

$$C_{L\lambda} = C_L/\cos^2\lambda \qquad (343)$$

Utilizing Equations 338, 341, and 343, Equation 340 becomes

$$\text{(Swept)} \ \sigma_{cr} = 2.45 \ t + 0.56 \ C_L \qquad (344)$$

Consequently, the cavitation number of the swept hydrofoil given by Equation 344 is exactly equal to the cavitation number of the unswept hydrofoil given by Equation 339. Therefore, no cavitation advantage is obtained from sweepback when a hydrofoil is strength limited.

APPENDIX E

CAVITY DRAG COEFFICIENTS FOR HYDROFOIL CROSS SECTIONS WHICH CONSIST OF TRUNCATED ELLIPSES

The cavity drag coefficient of a hydrofoil with a base cavity is given by Equation 242 as

$$C_{dc} = C_{do} + \frac{\frac{\pi}{4} (\sigma \ t/c)^2}{\sigma \frac{t}{c} + 1.5 \ C_{do}}$$

where an approximate expression for C_{dc} for the special case when $\sigma \ t/c > 4 \ C_{do}$, is

$$C_{dc} \doteq \frac{\pi}{4} \sigma \frac{t}{c} \ (\text{for } \sigma \frac{t}{c} > 4 \ C_{do}) \qquad (345)$$

C_{do} is the value of C_{dc} when $\sigma = 0$, assuming that the hydrofoil sides are wetted and the base cavity is at $\sigma = 0$. This case is mathematically possible to analyze for truncated ellipses, but it is not physically realistic because the surface of all ellipses will cavitate at $\sigma = 0$. However, if C_{do} is calculated using a mathematical approach which assumes that the sides are wetted, Equation 242 will provide a value for C_{dc} which is both mathematically and physically valid for all operating situations where $\sigma \geq \sigma_{cr}$, where σ_{cr} is the incipient cavitation number of a given truncated elliptical strut.

The value of C_{d_o} for truncated ellipses can be calculated from an expression developed by Tulin (27) using linearized theory for the case of $\sigma = 0$. This expression, in modified form, is

$$\begin{pmatrix} \text{Truncated} \\ \text{ellipse} \\ \sigma = 0 \end{pmatrix} \qquad C_{d_o} = \frac{2}{\pi c}\left[\int_o^c \left(\frac{dy}{dx}\right)\frac{dx}{\sqrt{c-x}}\right]^2 \qquad (346)$$

where y is the local semi-thickness and x is the distance from the leading edge. Equation 346 is based on the assumptions that: (a) the two strut surfaces are wetted, (b) only the base is covered by a cavity, (c) $\sigma = 0$, and (d) the cavity walls do not intersect.

Equation 346 can be placed in the form of an elliptic integral by nondimensionalizing it by substituting $x' = x/c$ and $y' = y/c$, and then letting $x' = \sin^2\phi$. Equation 216 is needed to express y' in terms of x'. After making these substitutions and letting $k = \sqrt{c/\ell_c}$, Equation 346 becomes

$$C_{d_o} = \frac{2}{\pi}\left(\frac{t_c}{c}\right)^2 k^2 \left[\int_o^{\pi/2}\frac{(1-2k^2\sin^2\phi)d\phi}{\sqrt{1-k^2\sin^2\phi}}\right]^2 \qquad (347)$$

where t_c/c is obtained from Equation 212.

The symbol k is normally used to designate the variable in complete elliptic functions which are usually symbolized by $K(k)$ for functions of the first kind and by $E(k)$ for functions of the second kind. From Equation 214, it is seen that k^2 is

$$k^2 = \frac{c}{\ell_c} = \begin{cases} \dfrac{\sigma_o}{2} \sqrt{\dfrac{C_1}{M'}} & \left(\begin{array}{c} \text{Region IIa} \\ \frac{1}{2} \leq \frac{c}{\ell_c} \leq 1 \end{array}\right) \\[2em] \dfrac{1}{\dfrac{M'}{\sigma_o^2 C_1} + 1} & \left(\begin{array}{c} \text{Region IIb} \\ 0 \leq \frac{c}{\ell_c} \leq \frac{1}{2} \end{array}\right) \end{cases} \tag{348}$$

Using elliptic integrals, the solution to Equation 341 is

$$\left(\begin{array}{c} \text{Truncated} \\ \text{ellipse} \\ \sigma = 0 \end{array}\right) \quad C_{do} = \begin{cases} \dfrac{\sigma_o}{\pi} \sqrt{\dfrac{M'}{C_1}} \left[2E(k) - K(k)\right]^2 & \text{(Region IIa)} \\[2em] \dfrac{1}{2\pi} \left(\sigma_o^2 + \dfrac{M'}{C_1}\right) \left[2E(k) - K(k)\right]^2 & \text{(Region IIb)} \end{cases} \tag{349}$$

where C_1 is obtained from Figure 32, k^2 from Equation 348, and $E(k)$ and $K(k)$ from standard mathematical tables. Notice that the two expressions in Equation 349 agree along the Region IIa to Region IIb boundary given by Equation 288 where $\sigma_o = 3.19 \sqrt{M'}$ and $C_1 = 0.098$.

The expression for C_{do}/M', obtained from Equation 349 is plotted in the lower graph of Figure 42 as a function of σ_o^2/M'. Notice that the value of σ_o^2/M' is graphed only up to 24.4. This is the value where $C_{do} = 0$ which corresponds to the physical situation where the two cavity walls just begin to meet downstream of the strut base. At values greater than $\sigma_o^2/M' = 24.4$, the cavity walls cross and the theory is no longer valid. The upper graph of Figure 42 contains a plot of the corresponding values of $(t/c)/\sqrt{M'}$

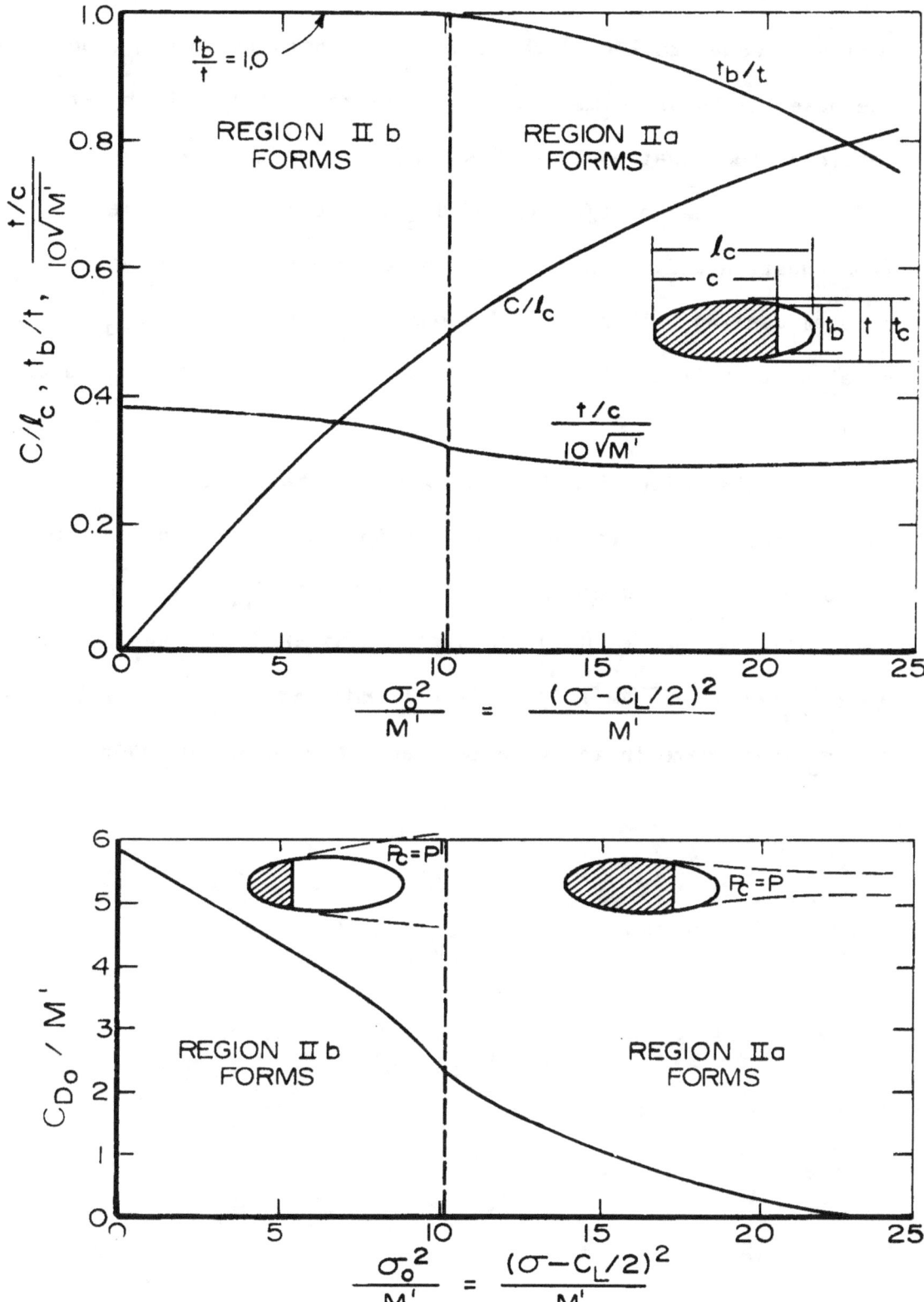

Figure 42 — Drag coefficients and physical properties of truncated ellipses

(which is equal to $1/\sqrt{C_1}$), the truncation chord ratio c/ℓ_c, and the base thickness to maximum thickness ratio t_b/t, which were obtained from Equations 212, 214, and 211, and Figure 32.

The values of t/c, σ, and C_{do} are needed to calculate C_{dc} from Equation 242. Since σ, M', and C_L are given in a typical problem dealing with hydrofoil cross sections, t/c and C_{do} can be obtained from Figure 42 using σ_o^2/M', where $\sigma_o = \sigma - C_L/2$ (Equation 294).

If the value of c/ℓ_c is between one-half and one, as in the case of the Region IIa hydrofoils of Appendix B then Equation 345 can be used to calculate C_{dc} because $\sigma\, t/c > 4C_{do}$. The latter inequality is seen to hold since Figure 42 shows for Region IIa that $\sigma_o\, t/c \geq (\sqrt{10.2M'})(3.4\sqrt{M'}) = 10.8M'$ and $4C_{do} \leq 3.4 M'$; also, $\sigma \geq \sigma_o$ everywhere in the selected section of mission space.